Correspondences

To Marquis, Davis & Hazlewood

Doug Rucker - Architect

Correspondences to Marquis, Davis & Hazlewood

By Doug Rucker
Layout by Helane Freeman

Copyright © 2025 Doug Rucker
All rights reserved

Doug Rucker
Vilimapubco
Malibu, California
ruckerdoug@gmail.com

No part of this publication may be reproduced, distributed or transmitted in any form or by any means, including photocopying, recording, or other electronic or mechanical methods, without the prior written permission of the publisher, except in the case of brief quotations embodied in critical reviews and certain other noncommercial uses permitted by copyright law.

For permission requests, sales to U.S. bookstores and wholesalers, or to inquire about quantity discounts, please contact the publisher at the email address above.

Library of Congress Control Number: 2025911804

ISBN — 978-1-7354717-9-2

First Edition
10 9 8 7 6 5 4 3 2 1

Printed in the United States of America

Correspondences

To Marquis, Davis & Hazlewood

JOHN MARQUIS FORWARD

John was 5'-11", 190 lbs. and unusually handsome with wavy brown hair and large blue eyes with long lashes. His natural skin was a deep tan

Bottom - Leroy Scheck, John Marquis, Doug Rucker. Middle - Unknown. Top - Bernie Simonson.

and he had a virile beard he kept neatly shaven. Rather somewhat slimmer, he reminded me of a young John Barrymore.

He also smoked and would inhale a long powerful puff, the kind that would kill me instantly, then exhale it through his lips only to inhale the same smoke again through his nose and then finally discharge it in a faded and thoroughly obliterated cloud. How he loved to smoke and how I hated to watch!

Smoking was prevalent during the forty's and fifty's. Many of my friends, including Mother and Dad and all my relatives, smoked. They had done this during my entire lifetime and I had always hated it. I was used to breathing second hand smoke, but even so fought myself trying not to judge the complete worth of a person by whether or not they smoked. I had to work at not be a prude. Today, I still hate smoking. Even if it means being a prude, I forgive myself for judging others on the basis of smoking. It's bad for you! At my advanced age I just won't put up with it anymore.

Coming from a well educated and rather well to do family both caring and loving, John was easy to like. His family was educated, straight and religious. He greatly admired his older brother, a newly married preacher who always did everything right. John was the black sheep. His major was Psychology and a good portion of his philosophy seemed to be based on *mind over matter*.

Lucky Strike John, evil ringmaster, with quartet and girls of the night.

John was also a bit of a night owl and didn't seem to need much sleep. He was interested in people, books, partying, beer, carousing and conversation. I was a morning person, interested in early sunlight, fresh breezes, nature and the outdoors. In the late evening, about the time I'd be crawling sleepily into my lower bunk, John would come in from the great outdoors and decide to read what seemed like an entire book in the upper bunk and smoke himself to death. The light was always in my eyes. I could rarely sleep with the light on and the putrid, eye-burning smell. How I suffered! I told myself I'd eventually make a fine marital partner because I'd have had practice putting up with another's bad habits and impossible idiosyncrasies.

John's vocabulary was immense and acquired from great quantities of reading. He enjoyed being well educated and used his natural curiosity within the framework of encouraging and loving parents. Sometimes he would use a big word and I'd ask him what that meant and he'd define it for me. He used big words in many social situations and didn't apologize. He wasn't bothered by those less talented verbally not to be understood. I was always asking him what he meant by this or that and he would always graciously reply. I am more of a reader now, but still impressed by his remarkable and continuing ability.

After 45 years when John and Pat were still living in Palo Alto with their three well-educated and professional children, non-smoking John and I were both graying and still friends. Since then, he and his wife have passed into the great beyond. What follows are letters from me to John Marquis that were fortunately still retained on my computer. Sorrowfully, John's returns to me have long since disappeared. Our letter conversations were during ten years between 2003 and 2013.

4-9-03

John and Pat,

   Though we have not been in touch very much for the past part of our lives, I wanted you to know that I have always thought of you with warmth in my heart. We were roommates together during a special time in our lives. I remember us in college playing in the Intramural football games, our intelligent discussions, your use of big words I'd never heard before, but that made me feel important because you didn't lower your standards for me and always explained what the words meant with no apparent irritation. I loved our discussions about your preacher brother, father and mother whom you loved and respected so dearly and your unique and educated, sense of humor. *(I was hard put to it because of your smoking and late hours, however, so you were not perfect. I of course, was.)* I'll forgive you if you'll do the same for me. I admit I had a few problems of my own.

   I am sending you a poetry book I recently compiled about work I did that was written between 1966 and 1984. *(18 years)* It is called *Moving Through*. Please read the forward and the back cover of this new book originally made by hand. I wrote all these poems while I was moving through my divorce period and when I'd finished *Moving Through,* The title occurred to me, *Moving Through.* I quit attempting poetry for almost 19 count 'em, *nineteen* years. I had written 13 small comedy books in the previous 10 years before my stroke in 1992. *(For quite some time, perhaps a year or so, I had trouble verbally communicating — I could understand what was being said and knew my reply, but was extremely poor at phrasing new conversations.*

At the same time the stroke diminished my desire to be humorous changed in the same *off the cuff* way. My sense of how to be funny wasn't gone, but remained modified. I couldn't especially think of anything ridiculous to write about, nor did I *want* to think of anything ridiculous to write about. I didn't lose my sense of humor, but felt my attention should be directed to something of more importance. *(I began the autobiographies.)* Maybe this was where I could use my humor.

I'm still busy doing architectural work at the age of 75 and have just finished 1-1/2 years doing the *Carolyn Van Horn* additions. Off and on, Carolyn was the Mayor of Malibu and City Council Member over the last ten years who is now widowed and retired. Until she died, there was a complete living quarters for her mother over an existing garage. *(We tore down the original room and garage because it was easier to start over than to work around the old construction.)* The new room would be for Carolyn. It came out wonderfully well. The new upstairs addition has glass on 4 sides. The South side views the open ocean with the island of Catalina. The East side looks over the entire Santa Monica Bay ending in Palos Verdes. *(See map. Carolyn lives on Point Dume.)* The northerly side views the Santa Monica Mountains and the West side — close-up views of foliage, fat trunks and tall pine trees.

I'm also doing some landscaping for *Steve Wolf* consisting of a large outdoor gathering space and dance area for teenagers, a bar-b-que with seating for four having soft drinks, a sitting area around an out-door fireplace, a rock pool/fountain and flower garden outside the new dining room I designed for them 2 years ago. *(I had done a two-bedroom, bath, hallway and home working studio for two about 6 years ago and had done the original house about 20 years ago.)* The house looks better today than it did when it was first built on a sloping one-acre lot under a canopy of live-oak trees

I'm also re-doing some decks and bridge, retaining walls and a new driveway to a house I'd done 25 years ago in the Western part of Malibu for the original owner, Joe Cleary (81), former owner of the Malibu Lumber Company and the Fischer Lumber Companies in Santa Monica.

I have just finished doing a new $25,000.00 driveway for the Gillin's who were former clients and owners of one of the first houses I designed in Malibu, 30-years ago. It came out very well.

Then, I'm helping eldest daughter Viveka and her fiancée, Tom Rincker, with some of their additions. I just finished measuring and transcribing their house and entire property. *(Lot- 64' x 160')* That job will proceed slowly as their money comes in slowly.

Marge has two art pieces in an art show devoted to former Art Therapy graduates of Loyola University. We attended the opening last weekend and her drawings *(she does meticulous 18"x30" PrismaColor drawings)* I thought were some of the best in the show. She has completed about 15 drawings of the same size depicting items in her studio that remind her of certain episodes in her life. She draws various items her memories inspired. If and when she prints them, I'll send some along to give you an idea.

Our kids are doing well and are in the throngs of their lives, as anyone would be in their early 40"s. Their lives would take another long letter.

Regarding my spare time, and I use the words spare time loosely, my hobby is making books. My philosophy at this regenerative part of my life is to produce something tangible to leave behind. These include stories of my life and how it was to live in our age, my thoughts on life and living, my artwork, if any, my architectural work *(to be included in volume 3, 4, 5 of my autobiography)* and some indication of what kind of guy I was in the age of our lifetimes.

I tell about me, but I also tell about the persons who were close to me and who shaped me and formed the next part of my life, that were hopefully the good part. I decided, before I began Volume 3, to put some poetically inspired thoughts into a book that would help to explain whatever the book explains. *(Read the back of the book first, then the last selections, cover and forward.)* I am honored to share this book with you. The poems are almost 20 years old, but still have special significance. *(Perhaps to you, too.)* I have a poem called *He Leaps* which is about a cat that leaps to a distant table, misses, scrapes his claws and falls in disarray to the floor. The final line is. *"Until I remember, he did leap."* I consider I'm like the poetic cat *I did leap.* Hope you enjoy it.

    Best wishes, and keep in touch.
            Doug

3-30-05

Dear John and Pat,

  Regarding my autobiographies, I finally finished volume 3 following the mildly adventurous saga of an Illinois architectural graduate whose roommate was the legendary, barrel clad, John Marquis. It has been a long time and a lot of water has passed under the bridge. The water that's passed is what the book is all about. Am I living in the past? Well, not exactly.

  I've submitted the preliminary drawings for the second time on the Kris Kristofferson job in Malibu. Since the Coastal permits are now handled by Malibu City, former applications must be reviewed with restrictive *Coastal Commission* rules in mind. I think I told you, it's a 2,000 sq. ft. house on a hill overlooking magnificent westerly views of *Surfrider's* beach, *Latigo Point* and *Point Dume*. I have just gotten Geology and Health department approvals that besides *Coastal Commission* approval are the toughest agencies to conquer. Of course, if the geology on the Kristofferson lot fails and the site slowly begins to creep down the hill with these record-breaking Southern California rains, it will also include another of my designs - the 6,000 sq. ft. McCarty house. We don't want that to happen, do we?

I'm also doing a kitchen for a former client *(Darlene Beaver)* in Brentwood Village. It's a small job *($50,000.00 or so)* in a very old house in an exclusive location. The building is worth nothing; the half-acre lot is worth two million. Oh well! It's for another Psychologist with a booming Santa Monica practice and you can never tell what they'll do.

Then again, I'm designing a third house for Mike Bright, a surfer and Olympic Volleyball Champion. He once knee-paddled from Manhattan Beach to Catalina Island and back *(32 miles)* and won the race with the nearest prone-paddler coming in 5-1/2 hours afterward. Is this something to admire or what? Unfortunately about ten years ago, due to faulty equipment while diving to check lobster traps, he was forced to ascend too fast from a 150-foot depth. He couldn't get to a decompression chamber fast enough, got the bends and lost the use of his legs. Last September his younger wife died of congenital heart failure. He'd bought the burnout lot a few years back for a buffer, but now will be custom-designed for a wheel chair patient. I've been working on it today. I'm enthused. It's an exciting challenge.

Between architectural jobs, I work on my computer telling stories. I'm not bragging. I've been reading many books by Wallace Stegner and John Steinbeck and they keep me in touch with the reality of writing.

So, though I live sometimes in the past, I also live in the present. I'm making a good past to tell stories about later. *(That is, if I'm still alive.)*

Marge is fine, going to work as a grief counselor, going to an exercise program three times a week called *Curves*, doing Tai Chi once a week and running two miles on Saturday. For a 71-year-old lady, I think that's amazing.

The kids are fine except my youngest daughter, Amanda, is in the throws of a divorce. She is a single working mom taking care of two brilliant and energetic boys, Nathan, 5 and Christopher, 6.

I'd love to hear from you. So far, we're still alive and must be appreciative for each day. We'd welcome a visit from you and your wife at any time. If we are up your way, we'll call and stop in. Wishing you health, happiness, passion and intelligence. *(The last two, I guess, you already have.)*

Your roommate,
Doug

2-23-12

Dear John,

   Got your call yesterday afternoon. It was great to talk to you. Looking back on our conversation, it sounds like you've been to death's door and back. I am so sorry that happened to you. Might I even say, again? You know. You and I are further bonded because we both have pacemakers? Lucky you! I had another old college friend. *(A veteran)* He was glad to be almost ninety, but that was just *"considering the alternatives."* I will be 85 in 8 more days and I'm feeling it in many parts of my body; hips, knees, breathing, etc. In the past I've seen elderly men barely shuffling, with their shoulders bent, poor eyesight, wearing hearing aids, gray hair or balding. Now I know why; they HURT!

   I feel I'm doing well for my age. Fortunately, my mind is either strong or I'm terribly deluded, but my hips hurt when I go for a walk. Particularly walking uphill. Then, if I sleep on my back wrong, when I get up it refuses to straighten out no matter how much pain I'm willing to endure. I have to take a hot shower before it limbers a little and I can get out and walk. After walking a while it seems to do better.

A year ago I was surprised to find a shallow bulge of water on the side of my left knee and could marginally bend it. It was water on the knee. I checked with the doctor. He said he hoped for the best and I was released with the vague hope it would go away by itself. They say the body is the best healer. I hope they are right! I'm still able to go for a walk five mornings a week. I'm incapable of jogging because my hips and knees hurt too much. I go walking for about 50 minutes with knee and hips hurting every step. After a year, I think I'm seeing a little knee improvement.

Also, a little over a year ago on Halloween morning, I tried to climb over a four and a half foot chain link fence and my heel hit the top as I tried to jump to the ground and instead I fell like a tree on my back and broke a rib that punctured a lung. I also got a humongous contusion on my elbow. After a year, I'm substantially over it, but had to quit early this afternoon to rest from the yet residual pain. By the way, I'll never take Vicodin again; the drug gave me constipation and mind-alteration.

Then about five years ago I almost died from an infection on my left elbow. It developed from a scratch I thought had healed under a drafting callous on the point of my left elbow. I went to the emergency hospital and stayed there for a week with a flood of all-purpose antibiotics pouring into my vein from a bag on a pole. Then, when they sent me home I had to remain inactive with the bag and pole for two more weeks. The emergency doctor who'd been the UCLA infection specialist for 13 years told me the last time he'd seen an infection like mine it was on a woman who had flesh-eating disease. He said they could only save her life by removing all the skin from wrist to shoulder. This they did and I presume she is still alive. It looks like modern science prevented me from death by flesh-eating disease; all this on top of a pacemaker implant I had ten years ago. I lived through the intelligence and good will *(and Federal money)* of modern medicine.

All this is to let you know you are not alone. And still, I'd rather have gone through what I went through than spend time, as I suspect you did, in the hospital with it's anxiety, pain, discomfort, bathroom and washing difficulties, injections and intra-feeding and total change of environment I'm sure can take the stuffing right out of anyone.

I'm surprised humans can withstand what they do. Humans must have a fanatical desire to live in order to put up with it. Lately, I'm sort of familiar with hospitals through Marge's recent experiences. A year ago she was diagnosed with Diabetes 2 and has been treating that moderately with twice daily blood tests, diet, pills and carrying a supply of emergency food wherever she goes. Then she developed an involuntary movement of her hands, perhaps essential tremors. The doctor requested a brain CAT scan and a UCLA neurologist diagnosed her condition. He thinks she has early Parkinson's disease, for which there may be hope for at least partial cure - or it may be essential tremors, for which there is no cure.

Then Marge noticed a swelling along the insides of her groin - where the thigh meets the body - and the doctor diagnosed it as swelling of glands and ordered a full body PET scan. It revealed the glands in her whole body lit up like neon tubes. A week ago she came out of surgery having had an entire gland removed from under her left arm for a biopsy. *(There's always a good side and always a bad side. If a person loses a leg the bad side is that he lost the leg. The good side is that he didn't lose both legs.)* The follow up is that she has a low grade lymphoma that will always be with her. On December 10th the oncologist will be digging into the bone of her hip for a bone marrow test to prove that's the origin of the lymphoma. If that's the case, they'll follow up with eight, once-a-week injections of a milder chemotherapy. Theoretically she will not lose her hair, but her immune system will be heavily compromised. She is presently affected by fatigue and exhaustion and has to take long afternoon naps. *(So do I, by the way - but not so long.)* Surviving chemo, the prize and hope is: the lymphoma will go into remission and she'll regain her former energy.

So! The way to make you feel better is to tell you all *my* troubles. John, my hearing is so bad. I'm wearing two hearing aids, and though I think I heard everything you said over the phone, nevertheless, I'd appreciate it if you could write me a letter and tell me about your health and procedures and anything else I might not have heard, so I'll fully understand. Are you really on the mend? Will there be recurrences? The thought of dialysis scares me. You were a great roommate and are a valuable person. I don't want to lose another friend. I'm running out! You're still alive! Write soon. I can't wait to see your crazy, but loveable handwriting. My best to you and give my best to Pat. I know what a strain these complications can be for you and her.

Your friend and roommate,
Doug

3-28-13

Dear Doug,

   Sorry that you have pain & health problems that make things hard. I am on dialysis now 3 x a week for 3-1/2 hours and got 1/2 way through Cookies? Especially enjoyed the description of the walk in your little heaven environment and time Marge saw the TARANTULA!

   Dialysis is making me feel better and I'm getting a little of my old spring euphoria & getting eye exams for my dental work done. I'm enjoying my time on the machine reading & listening to book on Ipad *(MT's Innocents Abroad)*

   Pat's health is good except stable eye problems. Kids doing well & happy with relationships, except Neils's wife Janine has Guillon-Baret Syndrome, but is recovering well.

   We still host family parties. *(Our house, but kids do all the work.)* It's hard for us to travel now, but we are enjoying still being in our home & being retired.

I am eternally grateful for having your old autobiographical & geographical books to enjoy when I was in the chemo room when I had a dose of cancer.

>Much love & thanks for the Cookies.
>*John & Pat*

4-2-13

Dear John,

    Thanks for your letter. Glad *Where's the Cookie's At* is keeping you company. Marge was particularly delighted you liked the *TARANTULA* story. I was going to toss it, but needed a couple of extra pages.

    I was shocked to hear you are on dialysis three times a week for 3-1/2 hours. Having your blood removed, cleaned and replaced has got to be tough on you, but I'm glad you're feeling better. It's good, too, to get caught up on your eyes and teeth. Good for you! I don't think I knew you had a bout with cancer. I guess that's under control. I've had cataracts removed in both eyes, have a pacemaker, my hips hurt when I walk and am on pills for high blood pressure and high cholesterol. Oh yeah! I'm also 85.

    I'm not familiar with *Innocent's Abroad*. Is that a book or program or series or what? I'm also glad your kids and their relationships are working out. Janine's Guillian Barre' Syndrome doesn't sound good. It's a great sign if she's recovering.

I'm glad you're enjoying my autobiographical books. I think I sent you *Early Stories, Groundwork and Growing Edge*. I've finished 267 pages of book four that I think I'll try to publish even though it's not finished. I've kind of run out of gas and am not sure I'm going to do any more. I think I'll call it *Malibu Practice*. I was 42 in those days and it covers the aftermath of the fire that burned the pedestal house to the ground in September of 1970. That book completes a major chapter in my life. I didn't meet Marge until 1976 and that would be the next chapter.

Marge is on chemo treatments for lymphoma, a cancer of the lymph glands, this on top of her diabetes and Parkinson's disease. Including two PET scans to spot the problems in her lymph glands, she's also had an MRI to locate any aberrations in her brain. I'm glad to say none were found, though that doesn't keep her hands from shaking. Marge is doing well having just taken her 5th infusion of Rituxan - a drug that only fights the bad guys and leaves the good guys alone. What this means is that she feels pretty good after the infusion and isn't nauseous and doesn't lose her hair.

What's almost as hard is the fact she has the

beginnings of Parkinson's disease. Her hands want to shake all the time, an especially difficult problem for a PrismaColor Pencil artist. When she's awake, they shake, but when she's sleeping, they stop. When she tries to relax she can sometimes stop the shaking. Hands, shaking is difficult, but she's stoic. If you met her right now, I doubt you'd know she was having physical problems. Despite her long naps in the afternoon, she pulls off social Gallery openings quiet well. She's now retired from being a Grief Counselor at the Thousand Oaks Hospice of the Conejo. Conejo means rabbit in Spanish. I don't know how that came about. When she's not doing Doctor stuff, she works on taxes and bookkeeping and bills and goes shopping and does cooking and is a marvelous companion. I'm madly in love with her.

Now about ME! After a layoff of about 3 years, I fell back into the architectural habit. The old bell went off and I dashed into the ring and started swingin'. Also, it seems my past has come back to haunt me. When one or two of my houses recently came up for sale, Malibu Agents advertised my houses as Doug Rucker Malibu Classics. Apparently, I have done *"Classic"* houses. One job I'm presently doing is an addition to a house I did 53 years ago that still looks *great!* The owner built it in 1960 for 30 thousand dollars and the owner just bought it for a million-three. Where does an alien go to register? The other job is trying to make a new house out of a 2400 sq. ft. barn-like structure.

Also, I'm continuing in my reflective photography artwork and am in a couple of shows, 4 pieces in the Ventura Atrium Gallery and 5 pieces in the Ventura Community Memorial Hospital with two more shows coming up. Also, I've just finished another colored book of 70 pages with 25 pictures of my reflection photographs. It's called *Reflections*.

My kids are still talkin' to me, so I guess our relationships are OK. They're doing quite well. Married to a computer man, Viveka, my eldest at 52, has her own flourishing Acupuncture business. My 51 year-old married daughter, Lilianne, with three children, is a Practitioner for the Center for Spiritual Living near Santa Rosa and drives emergency hospital blood. My youngest, Amanda, at 48 is a divorced mom with two boys, 10 and 12. She works full time in Malibu as a Registered Nurse doing home health care in Malibu.

And that's some of the news from Lake Woebegone. Thanks for your letter. It was good to see your tangled handwriting. I'm used to it! Give my love to Pat and send me a note when you get a chance.

Your old roommate,
Doug

CHARLIE DAVIS

CHARLIE DAVIS FORWARD

In college I made friends with Charlie Davis, an Architecture student. In about 2014, he and his wife, Lucille, died a month after an automobile crash. The couple lived lifetimes after college in Roanoke Rapids, North Carolina. When they died, Charlie was 95 and his wife a bit younger.

Friend and Architect, Charlie Davis.

We shared the following experiences in architecture class: our major and principal class at The University of Illinois between 1945 and 1949. Class was 4 hours a day, 3 days a week. The lights could be seen blazing from the upper windows of our Architecture building until midnight when the building closed. Students were using every ounce of energy finishing projects exhibiting dedication to their chose subjects.

First year students were told the Illinois Architectural School had one of the best reputations in the USA. We were under the *Beaux Arts System of Design*, a method selected by most of the U. S. major Universities. All given the same projects at the same time as competitions that won prizes for Excellency. It was common knowledge that Illinois had won ten first prizes in the former ten years. Asian students seemed to be the best. They worked hard and had an ability to draw unlike American colleagues.

I remember looking at the Oriental seniors work and thinking I had no chance of being as dedicated or as talented as they. Some were legendary. I don't know what happened to them, but I suppose they became excellent architects and did important work.

There was camaraderie in the drafting room. Throughout the large room, Sophomores, Juniors and Seniors worked together on their special projects in specified places. Because we were always there we knew each other and usually liked each other. Classical music was played softly by someone's radio and compositions by *Beethoven, Stravinski, Bach,* or *Shoshtokovich,* filled the room with inspiration that complemented creativity and gave us a sense of being at home. When more contemporary music was playing, it was music like, *How are Things in Glackamora, I'll Walk Alone, Long Ago and Far Away or I've Heard That Song Before*. There was no rock and roll, rap or un-sing-able songs.

I loved my board and was pretty good at my projects. I worked hard and liked soaking heavy Strathmore watercolor paper stapled to my board and returning to see it ready for my new designs. I was a fair draftsman and I loved my tools, pencils, tracing paper, brushes, inks, T-square, triangles, French curves, lettering guide, compass, etc. Though not visualizing myself as an Architect, yet standing at my table or fussing with my projects gave me a sense of purpose. My dad also loved his tools. His model-making tools were always oiled, cleaned and sharpened and his table was well lit and ready for new work. So was his work. Was it in my genes or did I learn how to work from Dad?

At the end of a session, papers and materials would be in a working mess on my board. I'd have graphite on my hands and smudges on my face, and I'd be tired and looking forward to a rest. I straightened everything, cleaned my board and left until the next day for new projects. I still do this, and so did Charlie until graduation in early 1950.

Saved on my computer since almost the turn of the century are letters written from Doug to Charlie between 2001 and 2013. Letters from Charlie to Doug have disappeared and are no longer available for reprinting. Here are Doug's letters to Charlie.

5-30-01

Dear Charlie,

   I am listening to side two of the Davis organ recital. You are playing something by Bach, only the program shows it's *Joyful, We Adore Thee*. It's lovely. It originated from Bach's, *Christ lag en Todasbandan* or *Christ Lay In Death's Dark Sleep. (or Prison)*. We used to sing it when Karon and I were members of the Neo-Rennaisance Singers. Your organ playing sounds professional. I'm enjoying your playing. It is remarkable how well you're doing. Thanks for sending me the tape. Also, I would like some more Davis poetry. I have a Davis file that won't get lost.

   Also, I always enjoy pictures being sent. Pictures tell a thousand words. That's why I included so many in my book. Words are one thing, but never explain anything as well as a single picture.

   Today I called one of my old High School friends, Jim Baer. Jim and I played left and right halfback on the football team for Chicago's Austin High School. We also tied for third in 1941 and 1942 *(almost 60 years ago)* in two City pole vault championships and though the sands of time have changed, we both still weigh about the same. I sent Jim copies of my two books. *(a)* Because we have been close friends. *(b)* Because on the cover of *Early Stories*, that's Jim walking with me along the downtown streets of Chicago. It's my thought a person on the cover should have the book. Of course I sent him pictures of Marge and me, the best ones, of course, those that make us look young, vital and smiling - like we've just had a big slug of apple cider vinegar and honey.

I also sent a letter to Don McGregor, an old High School friend who lives in the Chicago area. He, as you, loves humor and so I sent him a bunch of silly stuff — the kind of which we have plenty here.

I am sorry for the mistake on our age difference, but you have to admit, you're looking so young, anyone could have made that mistake. Yes, I'm exactly 73-1/2 tomorrow. I was born on December 31, 1927. I can't believe you're 82. I see by the picture in your office it is one by Fiffle's, the one showing the railroad track scene. From here, it looks like an excellent painting. I presume in the standing scene behind you is a photo by Pallayo Lareno, or is it Kemp's work?

On the painting note, Marge has been reviewing her life through doing PrismaColor drawings of the items in her studio and the things that come to mind around them: rocks from the beach, souvenirs she collected and my attendance of the 17th century Renaissance Fairs given in Southern California each May and June, or pictures of her mother's artist table in her former home, and so on. She is quite busy with her artwork. Both her parents were fine commercial artists. She does this in her greenhouse studio which she also uses to see her three or four private clients. She is a licensed Marriage, Family, Child Councilor, *(MFCC)* and is entering her tenth year as the respected and beloved leader and counselor of those who have lost their spouses. She works for the Hospice Group of Conejo Valley, and has three groups, about fifteen in each. She derives deep satisfaction assisting grief-stricken people to reestablish their lives.

I continue to be about half-employed. That is to say there are always a few good jobs scurrying around out there. Occasionally one will take the bait and we'll establish a long friendship while making an excellent house.

At present I am doing work for the former

Mayor of Malibu, and a woman named Carolyn Van Horn. I would suspect she is in her late fifties or early sixties, has one thirty year old son who is legally blind and another married with two children who works for the Police Department. She apparently made an arrangement with her married son to live in and partially own her house while she makes an addition — a private one-room house over the garage. I have designed this for her and since she has sensational ocean and mountain views, is very pleased with a practically all-glass structure, including, if you can imagine it, an eighteen foot square skylight. I've warned her about the heat, but she knows and still wants it. We will use tinted double paned glass, with a Low-E factor to prevent ultra-violet rays entering. The skylight will also open at the top to allow the standard westerly trade winds to blow through the glass, up and out the top. She hopes, as I do, the building will be sufficiently flushed with air and comfortably cool. I doubt it, but it will be a good experiment as long as I have a back-up position.

When not dealing with clients or working with the Malibu Planning Department or some other bureaucracy, I do writing on the side. I enjoy it and feel it is something I need to do. Marge says

I am in the normal regenerative phase of my life, the time when a person feels he has something to pass on to the younger generation. I guess I have reached that stage. Between us, my kids seem only mildly interested in the books. I do have a few associates who have read and are reading them, and that gives me appropriate feedback. I'm writing the books to see what happened in my life. I expect I will have a better understanding of what life is about, the good and bad and things that can happen. I am reminded of the question, whether or not the unexamined life is worth living. In my work I expect to get a different perspective of myself, sort of look at my life more objectively; see the world and those who've occupied the world with me during my short time here. What I do with that, I'm not sure. Perhaps I will come forth with sage comments, or else quietly die, unexpressed. I'm sure it makes no difference one-way or the other.

I still have to figure out how to send and download e-mail pictures. Do you know? Thanks for the letter, the pictures and tape. I appreciate all. Please send more.

Sincerely,
Doug

11-10-01

Dear Charlie,

   Yes, indeed, there is something haywire in my e-mail. I can't even sign on. After the weekend I will talk to my son-in-law, Tom Rincker, the electronic-computer genius, and we'll see if he can fix it. I don't think it's much. I'm sure it's just a small glitch. I'll let you know either by normal mail, *(no white powder)* or by sending you a Gmail. Your address for me is correct, by the way. Thanks for telling me of your friend's experience with carrier's headquarters. If I need to, I'll call them.

   No I'm not swamped with architectural work. I have one job that seems to be methodically proceeding despite the Malibu Planning Commission's bureaucratic efforts to swamp a simple addition over a garage. In Malibu a good bureaucrat will do everything he or she can to make it more difficult to build. That's their job. At times I'm glad I'm forced to be semi-retired.

   I did have a couple of good prospects in my studio yesterday morning, however. They own a 2,400 sq. ft. condominium six miles away in Westlake Village, which they would sell, and an acre lot six miles away on the other side of our house near the ocean on Point Dume. Point Dume is becoming the playground of the *well-to-do.* They seem to have the money. He works as Head of Design and Construction at Cal-Poly Technical College in Pomona City, sixty miles inland. *(A long commute.)* He and his six-man staff provide the interface between school architects, contractors, and college administrators. The two new private college buildings he is working on each cost over three million dollars. A person in his position has immense authority and responsibility. His bright wife is one who edits corporate material, but she would prefer to edit books. Their children are grown and they have one grandson. They want a 4,000 sq. ft. house with garage and lap pool, which would cost between six and seven hundred thousand dollars over a three year period. I have to calculate my fee, which, at 12% would be almost $80,000.00. I gave the meeting my best shot by showing them photographs of my recent and older houses and invited them, if they'd like, to see them personally, I'm sure I am a professional person and I'm sure my jobs appeared to them to be professionally done, but I'm not sure they were into my style of contemporary design. They asked what my style was. I said *California*

Indigenous. Well, what the heck does that mean? I said I had a little squib I'd written explaining my style in more detail if they'd like to read it. She said, *"Yes, I'd like to read it." (I'll see if I can find it and send you a copy.)* When they left I felt a lack of enthusiasm about what I had to offer, no matter how well our professional meeting was completed. They were going to check a couple of other architects and when they had it narrowed to two, if I was one of the two, they'd contact me to see their lot and present living quarters for a more detailed meeting.

It was good to hear about your landscaping project, and I hope they didn't dig up your facilities. Showing up absent seems a good way to not attend meetings with the chairman of the Trustees. I've shown up absent on a few occasions myself.

It was interesting to locate your former best man. Had you seen him in 52 years? Sorry about his stroke. You know I had a mild stroke myself in 1992. It was in the lower section of the brain near the intersection of my spine and brain, the place where little is known of its function. They think it has to do with primitive responses, whatever that is. I suppose more ape-like. Afterward,

I had difficulty with phrasing sentences and using words. Resting and doing lots of reading, such as an 800-page history of Russia, helped me learn to speak more clearly. My difficulties in delivering the proper words and sentences stayed with me for almost a year.

In all other respects, I was fine. I could drive, eat, swallow, watch TV, read and looked normal. Though at first, tired, I could conduct myself in an ordinary way. I once attended an on-site meeting with a Johnny Carson-type person who did late night TV in Germany. He and his wife had bought a 3 million dollar house and property in Malibu to vacation on his summer off-season. It was located on the estuary of a little creek emptying through huge boulders into the wash and swirl of the Pacific Ocean adjacent to the shoreline property of *Whoopie Goldberg.* The General Contractor and Owner never knew I couldn't speak properly and I had decided to *"wing it."* Fortunately, the Owner was talkative and the Contractor so responsive, that if I was asked a question, I would lean cross-legged against a tree, finger resting casually on my chin, give a meaningful look and say '*Hmmmm!*" or nod my head affirmatively, or give a worried look and shake my head negatively. As you can see, I am all right now. Aren't I?

One more brief thing before finishing this visit, Marge's older sister, Barbara, and her husband, David talked Marge into saving for a cruise for Marge's birthday on the second of October. We would sail around Baja California to the little City of Loreto and back, stopping at La Paz and Cabo San Lucas. Due to a hurricane that had just swept Baja we were unable to visit there due to lack of docking facilities, but we all had a good time. It is the first cruise Marge and I have taken and we were delighted, not only with Loreto and La Paz, but with the accommodations on the ship itself. It was called, The Ryndam, a Swedish ship with Indonesian crew who was friendly and spoke to us in English. I'll send you a couple of pictures.

I'll let you know about my e-mail address when I get it straightened out. Wishing you and Lucille our best, I will sign off, now. Write soon! I enjoy the news from Roanoke Rapids. By the way, my friend, Don McGregor and his wife, Betty stayed in a cabin on top of Grandfather Mountain in North Carolina. Have you been there or do you know of it?

Sincerely,
Doug

12-9-02

Dear Charlie,

   I must congratulate you on the announcement by the First United Methodist Church of Roanoke Rapids that the chapel, designed by you in 1968, will now be named the Charles, C. Davis Chapel, and to recognize the 33 years you have given your services to maintain the structures of your design.

   Charlie, that's quite a feather in your cap and a real monument and appreciation for years to come. I think it's terrific and deserves the naming of the chapel in your honor. Do I get credit, too, because you are my friend? I can't appreciate you enough.

   Looking back, I think I sent you a Gmail in August. In June you had informed me of your daughter Ellen's death. It is my hope you and Lucille have found some way to live around that tragedy. Also, you had some personal problems like a tooth replacement, cataracts removed, colonoscopy and I notice you had diabetes. How has all that resolved? Do you now have a new and beautiful tooth, was the colonoscopy negative, are you taking something for the diabetes and are you seeing better now that you have new eye-lenses?

   I ask, because I'm entering the same time frame. I will be 75 on the thirty-first of next month. I have had a hearing aid, but lost it several months ago. In the meantime I've had a new examination and discovered I should have two hearing aids. Rumor has it the quality of sound depends on the quantity of money paid for the aid. I also have a prescription to have cataracts removed. Of course I am curious about that operation and want to talk to anyone who has had that done. Can you see any better? Will I still have to wear glasses? Can they correct astigmatism? In addition, I am scheduled to have two implants near my upper left molar. *(There's only* one natural tooth *left.)* Will it hurt? How can I get out of it? This year in September I had a pacemaker installed. I have had an irregular heartbeat and with all my athletics over a lifetime, my heartbeat is slow. This slowness with the irregularity finally caused several episodes of what they call *near-syncope (near-sink-o-pee),* which means *near fainting,* or almost fainting.

   The last time I fainted was at the University of Illinois bicycling home from my meal-job at twilight through the yet wild southern end of the campus. I'd picked up a two-cigarette sample package from the dinner table and, *no hands*, decided to light up. I did and deeply inhaled. The next thing I knew I was staring up at one bright star pinpointing through the deep blue sky. The twisted wheels on my bike had stopped turning, though I don't think I was out a full minute. That episode cured me from ever wanting to smoke again.

In any event, I could not afford to faint on the highway and risk crossing the line or banging into a light pole.

Enough of that! I'm still supervising my $360,000.00 addition over a garage and it's coming out beautifully. The plan is in the shape of a square 28 ft. x 28 ft. with glass on four sides, deck and 12-ft. sq. skylight. We just installed the windows and the cabinets are being made in the shop, however, the owner has plunked a mattress in the middle of the floor and is sleeping on plywood. I'm glad. She seems to love it. I'm also doing small work on houses I have designed in the past; driveway and more glass doors for *Gillin*, skylight and new roof for *Bowles-Carvalho* and expanding a bathroom and helping the garage to convert to a playroom. I have one good pie-in-the-sky job — two 3,000 sq. ft. houses — that I keep nourishing. I should sign a contract or not shortly after the beginning of next year.

In between jobs and on Saturday I'm continuing to put together a 450-page book of poems written between 1966 and 1984. It is sort of a *complete poetic works*, since I haven't written poetry since that time. Things came unglued then. My kids were doing funny things, My ex-wife, Karon, took two years to die of lung cancer and my new friend, Marge, got cancer of the breast and had it removed. I was busy!

At each of the 20-chapter headings, I will have a colored insert of a sort of art-therapy drawing. I'll send you a couple of practice copies. When I finish that book I plan to begin a book on the colored drawings, since I have about a hundred-fifty of them that I did in four months before I stopped. Tom Rincker, my almost son-in-law, who is a computer genius, though he wouldn't admit it, is helping me. For this he gets beans and weenies and gets to watch Monday Night Football with Marge and me. Marge does the cooking. Tom and I do the eating - but also I do the dish washing. Since it is almost time for Tom to show up, I'm going to cut this short.

Thank you for sending the notice about the Chapel that was named so deservedly and let me know about yours and Lucille's health and any other of your usually fascinating things.

To my lasting friend,
Doug

4-6-03

Dear Charlie,

   I received your letter of May 8, 2003 and felt badly the vision in your left eye was suddenly and so severely diminished. It must have been *(and must be)* quite a shock, having normal sight for 84 years then quickly finding your self in this condition. My hopes and prayers are with you that you will regain the sight in your left eye, or if that is impossible the Doctor's will find some way to make it easier to live each day. I know how sorry I would be if I had to lose my vision, not only because I love my sight so much, but also to give up driving. Time will tell and hopefully time will heal. In any case, you can see well enough to write and you must remember your friends love you and will be there for you, so we'll keep in touch. You'll keep us in the loop.

   I have been experiencing some discomfort around my heart lately. When I jog-walk after 25 minutes, suddenly I can't move. I'm stopped in my tracks. I feel tightness in my chest and I feel I have to belch. I don't hurt. I'm not short of breath. I just know I can't take another step and just stand there. After resting a minute or so, I can walk gingerly home. As a result, I have just had an angiogram. I think that's what they call it. They inject radioactive liquid into my arm, lay me down on a platform with arms over-head and move the platform and me through an MRI type of machine, which photographs the radioactive material flowing through the coronary arteries. They're checking for arterial slowing or blocking. Then they put me on a treadmill where I accelerate my heart to 123 beats per minute and at this peak number of beats per minute, give me another shot of radioactive material, this time to temporarily stain the inside of the coronary walls. After that I go back on the platform again and move through the machine again to have the stains photographed again. Maybe they used different colors. *(I don't know.)* Architecturally speaking, that's what I'd do. I have another appointment on May 30 to get the results and further advice. I hope I'm all right, but something must be causing the problem. We'll see. I'll keep you in the loop.

From a business standpoint I'm doing repair work on one job, replacing a driveway on another with retaining walls and re-doing a bridge and three decks on a third. On the Wolf job I'm still doing landscaping, an outdoor gathering and party-dance brick floor under a canopy of live

Oaks, a bar-b-que and teen-age bar and a fireplace and sitting area. On the Van Horn job I'm finishing a Mom's complete living quarters in one room over a garage. Including large entry, stairway, decks with glass handrails, granite counter-top and all Oak cabinets. It ran about $356,000.00. Then I have one new house I'm working on that aims to be building next summer. I'm as busy as I want to be. If I have extra time, I use it to continue with my autobiography. *(Volume 3)* Speaking of books, did you get a book of poetry from me I wrote entitled *Moving Through?* Just wondering.

That's about it from here. Charlie, Marge and I send you positive thoughts. Be well, write and let me know how things are doing with you,

Sincerely,
Doug

5-11-03

Dear Charlie,

I enjoyed your informative and humorous letter. Thank you for sending it. I love your letter writing - you tell everything that occurs to you and the devil take the hindmost. You and Marge are alike — wild-people.

I love your adventures with *up-front* Sunday's: members taking home public speakers, turning up the sound system for hearing impaired, former director swiping anthems, church never the right temperature, preachers first to break the rules. I've tried to imitate your style. Your letters are informative and give readers the real picture of your life and attitude.

Regarding giving you a painting of mine, I'll give that some thought and see what I can find. Hopefully you'd accept a full-sized copy. I have a few tucked away.

Regarding my supposed heart problem. The results of my angiogram were negative. The cardiologist, Doctor Yeatman, called me and used words like, *"good news, glad to tell… & saw no problems."* Marge and I met with him personally late last week, and though he indicated he wished my blood pressure a little lower, saw nothing to indicate heart problems. I was pleased by the news, but may still have to address what was causing tightness in the chest and stopping my jog-walk suddenly. I'll work with Dr. Baron on that.

I've been jog walking more slowly since I heard the good news and have not had any more symptoms. *(I don't take that news for granted.)* I have changed breathing while I run. I'm now allowing myself to breathe from the *bottom of my lungs*, instead of unconsciously taking a full breath. This allows any belching to occur which seems to be connected to tightness in my chest. As an explanation: if I exhale then exhale again more forcefully, that's what I mean when I say *breathing from the bottom of my lungs*. Or another example: I'll exhale, then relax and allow my lungs to fill when they're ready.

A traveling public health organization recently offered members of the community to test carotid and abdomen arteries for build-up of plaque using ultra-sound and photographs. This is to detect and prevent strokes. I should get results in two weeks. Since I had a mild stroke in

2003, I thought it a valuable test. My local GP, Dr. Baron, listened to my carotid arteries a year so ago and pronounced them fine. Of course I'm eating properly. Marge cooks using *The Zone Diet,* which works for us, and I *do* get 45-minutes of exercise with my daily jog walking. My weight is down. *(I weighed 151* pounds this morning).

By the way, regarding your arthritis problems. I am taking over-the-counter drugs called Glucosomine-Condroiton that after a few months has the ability to soften the membranes around the joints. If you are not doing this, do it. Also, I take 1000 to 2000 milligrams of MSM. After ten years or so, I can now snap my fingers. It feels so good. This prescription relieves arthritic pain. If you haven't done so already, check this out.

To bore you further with my health problems or benefits, I will be getting two new teeth, probably next month. I have had 2 new adjacent implants. That was quite exciting. I never thought I'd go ahead with it, but I did, and the implants look fine and are strong.

I kept an upper left back molar and had a bridge for years to the third tooth over, which was dead. The dead one broke away and I lived with a three-tooth bridge, bridging to nothing, a sort of cantilever for over a year, chewing carefully. Then I did the implant thing late in January this year.

So, other than having less and graying hair and ten or twelve subcutaneous but lumpy benign cysts and a pacemaker, I'm in pretty good shape. For a while I was undecided. I asked Marge, *"Should I get a Plasma TV or a pacemaker?"* She advised the pacemaker. When I get my implants, I'm thinking of having my teeth whitened and getting my cataracts removed. Then, when I look in the mirror, I won't recognize myself. I'll look young again, even if I don't feel so.

Marge and I are still working. With the little extra I've been making, we bought Marge a new $1,200.00 laptop computer. It has 40 gigabytes hard-drive and 512 mg memory. It will also play a DVD. We bought her a $200.00 printer with duplexer, plus the USB cables and a surcharge protector. She's now staying up nights to figure the darn thing out.

Also, with this letter under separate sheets, I'm including an event that took place last week.

I've sent it to the Malibu Times and The Surfside News. I hope they publish it. Marge and I live at the elevation of 1,700 feet and sometimes we're in the cloud. I'll send this along and write more later.

My love and, or at least best of luck to you and Lucille. I enjoy hearing from you.

Keep me on the list,
Doug

7-15-03

Dear Charlie,

Got your nice long letter and read it to Marge while she was cooking dinner. When she was finished, I was finished. She liked your letter a lot, as I too. You are accomplished and talented, not only in architecture, but when it comes to expressing yourself. You cover a lot of your life in 6 single-spaced typewritten pages.

The results of sonograms taken on my carotid arteries, both at the neck and the navel and the blood pressure tests I had on my elbows and shins, and my osteoporosis test taken on the heel of my right foot, proved I'm a good risk for a few more years. The photos of my left carotid artery showed it completely clear of plaque build-up. The right was very clear. Not perfect - but not dangerous. My navel showed minimal plaque build-up. I felt good about that test, because it showed that, at present, I'm not stroke-prone because of clogged arteries. My osteoporosis test showed I have none.

My tooth implants have been in for over 6 months and they say I'll be getting two new upper molars. I'm going to call Dr., Pfeiffer, now.

I did! They're going to do necessary appointment telephone calling. The implants can't be installed until late in the month because Marge and I will be gone from the 17th to the 24th. We're to fly to see Lili *(my 2nd daughter)* in Windsor, CA an hour and a half by car North of San Francisco. We'll rent a car at the airport and drive to Lili's where we'll stay a couple of days before taking a couple of days traveling through northern California to see Marrgy. *(Marge's youngest daughter)* She lives near Grant's Pass in Oregon. We'll stay with Marggy a couple of days then take a couple of days to return to San Francisco to fly home.

I'll be thrilled to get implants. My eyes have mild cataracts and they'll have to be removed before my next driving test, which will be sometime before December 31, 2005. We *have* to talk about what counts for us, don't we? *(Health, I mean.)* With implants, Pacemaker, new eyes and hearing aids I'm the new bionic man.

Since I've learned a new breathing technique, I have had no more episodes, which brought me to an abrupt halt when jog-walking. I'm going slower, primarily because my back and hips have been hurting me for about three months.

My back and hips were much better today, so I don't think I have arthritis. My pain is usually gone when I return from jogging. I'll have to get Viveka, my 41-year-old daughter who has just been licensed in the Nation and in California as an Acupuncturist. I hope I won't go under the needle. She is also a licensed Acupressure and Massage Therapist, so busy she has no time for me.

Regarding Glucosomine-Condroiton, it is usually available over the counter from places like Sav-On Drugs. If you have no large-scale drug place like mine at Sav-On, you should be able to get it at your local health-food store. Also there is something I've found that is even better than Glucosomine-Condroiton for joint pain, *(Glucosomine and Condroiton are usually taken together since one helps the other)* and that is a 500 to 1000-miligram pill *(or liquid)* named MSM. There was a wonderful actor (now dead) that swears he wouldn't have lived as long without the benefits of MSM. The MSM assuaged the pains of rheumatoid arthritis. My close friend and Illinois University trained architect, Don Patton, had the condition called rheumatoid arthritis.

Don Patton wrote his own 400 page autobiography entitled *My Life, Was It Worth It All?*, in the last seven months before he died at the age of 65. Don's leg was blown off below the knee by a cannon shell from a German tank while he was a U. S. paratrooper in Belgium. He was with the 101rst Airborne Paratrooper Division. After discharge with the purple heart he recuperated for a year and a half in the quiet lakes and forests of Canada just above Ely Minnesota. Afterward he attended Illinois in the architecture school and after graduating, married Nola Patton. The two had 5 beautiful children, while Don maintained a position as principal partner of a major architectural firm doing award-winning churches in Rockford, Illinois.

I got a glimpse of how terrible a disease rheumatoid arthritis can be. Don would have wished there was MSM when he was alive. I am using MSM myself and where for years I couldn't snap my thumbs normally, I can now snap my thumbs with no pain. Though rumor has it MSM doesn't work for all people, it worked beautifully for the actor and it works beautifully for me. I'll bet it will work for you. Try to get it. If you can't get Glucosomine—Condroiton and MSM, I'll buy some for you here and send it to you. You *must* have relief. How in the world can you play the organ?

On the home front, I've just finished building a shaded structure for our cars. At 1700 feet elevation in the Malibu hills, it's too hot. *We need shade.* No mosquitoes here, but we do have tiny black man-eating flies that buzz around the face and try for an opening in the eyes, ears, mouth, back of the knees, bottoms of elbows, anywhere there's human food. I went to the *Fence Factory,* where they sell fencing materials and bought $341.00 of 1-3/8" diameter galvanized fence pipe railing plus fence pipe connectors. It turns out the railings and connectors go together in a multitude of shapes like an erector set. I was able to build roof and sidewall frames to be covered with Sun-Brella, a long-lasting canvas material used for awnings. For the present it's covered with two $18.00 brown plastic tarpaulins.

When I'd finished the above, the next day we had a party. Marge had invited our gang over on Sunday. Chris, Marge's lawyer son and rock singer brought his fiancée, Diane, and her two teen-age daughters, Michele and ? They brought watermelon rinds filled with melon balls. Viveka, my oldest, *(43)* was there who did not bring her fiancée, Tom Rincker, *(he was working)* but did bring Mexican food. Scott and Amanda Jenson arrived with their two children, Christopher *(1-1/2)* and Nathan *(3)*. Amanda is my youngest daughter. *(38)* They brought brownies and ice cream. Jenny, Marge's oldest daughter *(46)* brought materials and made tortillas. We supplied — "the place".

We set up a picnic table and a dozen chairs under my new *(cheap)* carport, which has a sensational view overlooking the head of Bonsall Canyon, a fine view of Mitten rock and Calamigos Ranch and had a fine social time. The Lewi children, Chris, Jenny plus their two sisters, Marggy and Katy and their children had been on an Alaskan cruise paid for by Marge's former husband. The occasion was for us to learn about their trip. Just a side-note, but indicates an activity that goes on around here every couple of months or so. I love it.

Charlie, found a reasonable picture of Marge and me I thought you'd like to see it. It's enclosed.

Best Wishes,
Doug

11-4-03

Dear Charlie,

    I see I've been remiss in answering your good letter. My last one to you was in July, 2003. Your last one to me was September 1, 2003.

    First of all, did you ever think you'd live to see the year 2000? *(Neither did I)* I am thankful I got to see it. Soon comes 2004! I've lived 4-years past the millennium! To quote Charlie, *"Where does the time go?"*

    Regarding your July letter you said *don't be upset by computer mistakes.* Remember the adage *computers were made by humans.* And regarding you and Lucille listening to the Berlin Symphony Orchestra, I too, love the classics. Right now I'm listening to a CD of one of Beethoven's late string quartets.

    I enjoyed your anecdote about your la-de-da choir director who, several years back you invited to Thanksgiving. He was raving over Lucille's apple pie and you didn't tell him it was really Mrs. Smith's pie. Was that right? The director, as you describe him, *belongs* in a bank.

Apple pie reminds me of *Canterbury Farms Blueberry Jam* Marge received for her 70th birthday. She wasn't fast enough and I ate it all. Ah, whole blueberries floating in nothing but sugar and water. What a delightful waffle topping.

Also your information on the new preacher that allowed the congregation to request favorite hymns does seem strange, but it reminds me you were an accomplished soloist yourself and as I remember spent every Sunday in Champaign-Urbana singing solos at church and for weddings. I wish I had heard you.

My first wife, Karon, was an accomplished soloist. I loved to hear her sing solos as written by various classical composers. Karon had the ingénue lead in the musical, *Girl Crazy*, by George Gershwin. I was in the chorus and leader of the dance group. She was also a member of the Gilbert and Sullivan Society in Chicago before we met. One solo she loved to sing was in later life was:

If you're anxious for to shine in the high aesthetic line as a man of culture rare, you must get up all the germs of transcendental terms and plant them everywhere.

You must lie upon the daisies and discourse in novel phrases of your complicated state of mind.

Oh, the meaning doesn't matter if it's only idle chatter of a transcendental kind.

And every one will say, as you walk your mystic way, if this young man expresses himself in terms to deep for me.

Why, what a very singularly deep young man, this deep young man must be!.

This the first verse from *If You're Anxious for to Shine* from Gilbert & Sullivan's operetta, *Patience*.

We sang in the chorus together in the Neo-Renaissance Society and the Unitarian Church chorus for 9-years. In summer the Neo-Renaissance Singers sang at the *Pleasure Faire*, an immense pageant celebrated every year between May and June in the Los Angeles and San Francisco areas. We dressed in tights and gowns and sang works by *Morley and Wilbye*.

In the Neo-Renaissance Singers, I was listed as Douglas Rucker - *Bassus*. Others were listed as *Superius, Tenor, Superius-Altus, Altus-Tenor, Altus-Bassus — (or soprano, alto, tenor and bass.)* We sang works by Josquin des Prez, William Byrd, Couperin, Schutz, Weelkes, Gabrielli, Morley, Palestrina, Monteverdi, Dowland, Teleman, Vivaldi, Wilbye, etc.

It was ensemble singing at it's best and we had the best time. This music was a far cry from the music of the twenties my Dad sang me, such as *When You're Smilin', Where do you work — a John, When You Were Sweet Sixteen, Over There, Swanee, Rock-a-bye, Your Baby*, and the music of Al Jolson and Eddy Cantor. I learned all these songs including the songs of my own era and sang them to the kids, Viveka, Lilianne and Amanda when they were little.

I enjoyed your comments about your new preacher and I agree with you, putting up a screen with announcements on it and so forth. How about a bouncing ball over the words to hymns? If you can get over the tragedy, the situation is funny, or at least it's funny the way you tell it. There is much contrast in your characters. Don't apologize for ranting and raving. It's the essence of Charlie Davis.

Since we last talked, I now have my two new implants *(new teeth drilled into my upper jawbone)* They are working superbly and were worth the wait. Have you tried MSM and Glucosomine-Codroiton for pain and joint mobility caused by arthritis? Try it! You'll like it! My pacemaker has been working perfectly for the past year. This I have experienced and it is true even according to Sheila, the nurse who checks the instrument every six months. I have had no problems with my heart, even though I've gone for a jog-walk for 45 minutes every day except Sunday. I tried whitening my teeth too, with some stuff you paint on your teeth. I look better to myself. I'm not sure anyone else can tell the difference. It was wonderful! Last week when I told a stranger I was almost 76, he said, *"Boy! I thought you were only 63!"* Is there a message here?

I finished my tent-like carport cover we call *"umbrella"* It looks great and does the job — gets us out of the sun and helps during rain. See picture.

Since I've last talked to you Marge made a trip to Kauai to attend a three-day seminar in some phase of Psychology, a course in continuing education required for all licensed MFA'S. She stayed with her daughter, Katy and her husband, Marvin Otsuji and their 8-year-old daughter, Kainani. When Katy told Marge she was marrying Marvin, Marge said to herself, *"Oh! A nice Jewish boy."* Then she heard his last name, Otsuji, and discovered he was Japanese. So now they have a little Japanese-Hawaiian little girl aptly named Kainani.

When she returned she had a terrible bronchial cough that seemed to worsen over the weeks. She had to return to work and visit Bobby, her sister in San Diego because Bobby's husband is named, get this, *David Drucker*. The sisters married the Rucker's and the Drucker's. Who planned this? David, who has multiple Sclerosis was having an operation to remove a possibly cancerous tumor in his lower bowel. While he was doing that, Bobby had her other cataract removed. Why am I telling you this horrible stuff? Oh yes! Because all that studying, working and traveling was not helpful to her cough. Marge is fine now. Bobby can see. David is struggling and will be staying in a home away from home: great changes for the Drucker's.

During that time I have been doing my bread and butter jobs. A patio, BBQ and fireplace under a large canopy of oaks for Steve Wolf, repairing a bridge, decks and driveway for Joe Cleary, doing 2 rigid frame shear panels on a bluff overlooking 180 degrees of ocean and Zuma Beach, doing a kitchen and other alterations for Kim and Robin Belvin. Three of the jobs are finishing and the Eurich job is about to begin construction.

Last week Mike McCarty, a former client for whom I've done no less than five jobs for, called me. *(Mike is a restaurateur, is internationally known, has a West Coast establishment in Santa Monica an East Coast establishment in Manhattan.)*

He said the architectural writer for the New York Times, Joseph Giovannini, would be calling me. He was traveling to Southern California to witness the fires here. He did call and interviewed me by phone for 45 minutes. In that time I tried to relate all I knew about fires and rebuilding fire-resistant houses. You may remember, I lost 6 houses I'd previously designed in the 1993 Malibu-Topanga fires. When I hung up the phone, my brain was whirling. What came of it is the article I'm sending you. So, that's the story there!

The last item I must tell you about is the celebration we had for Marge's 70th birthday. Viveka, my oldest, and Jenny, Marge's oldest *(who are 43 and 46 respectively)* decided they'd throw a quasi-surprise party for Marge. Marge was to know there was a party but wouldn't know the theme. The theme was to be Greek-Roman. Everyone was to wear wreathes of fake grape vines around their heads and togas optional. They catered the party and we purchased drinks and favors. Kids were aplenty. Six of our grandchildren were there. Missing was Marggy, Marge's youngest daughter and her son Sammy and the Otsuji family from Kauai.

Marge was the Goddess and for her private entertainment, *(and everybody's)* Chris, Marge's only boy, Chris, *(42)* was master of ceremonies and gave a five-minute monologue on how great his Mom was. Then came Viveka, my oldest, and her fiancée, Tom, doing a rap-song. It was hilarious! Then, I had to say something and read her a love poem I'd written many years ago. Then two performers got up and sang her a duet with guitar. A special client I've known for years, Dan

Hillman, said another wonderful piece, while for the finale, members of her dance group danced an improvisational dance for her bringing her in to finish and gave her a group hug. It was fun and Marge is deserving.

That's all I can think of at this time, Charlie. Of course I was thinking about you all through the hurricane. Were you affected by it? It looked like you might be severely affected.

Best wishes and write soon,
Doug

7-12-04

Dear Charlie,

I got your e-mail. I tried to answer it the last part of June and got about a page done, then *it disappeared*. I tried to call it up again, as my daughter, Viveka, explained over the phone, under *"drafts,"* but I couldn't find *"drafts"* on my e-mail menu. At this point, I am still puzzled, however I didn't think it was such a good idea to risk losing a close and very dear friend for the likes of not being able to find *"drafts"* on my computer menu, hence this snail-mail letter. You might think, *"Why does Doug have e-mail at all."* It's because I hardly ever use the Internet. When I need the Internet, I have the Internet. Also, how would it look for a contemporary architect, doing houses for the rich and famous not to have Gmail and Internet? I'm a diamond in the rough coming late *(if at all)* to the new age.

I began a letter to you in the latter part of January. *(See enclosed.) Yes*! I now have my new computer. Tom assembled the main unit in less than an hour and a half and looked like a famous surgeon doing a heart-lung transplant and when finished, saying to the surgical nurse, *(me)* *"OK! Sew 'er up!"* Then, taking off his rubber gloves and white coat and leaving the hospital for golf, went home. Well, I suppose I exaggerate.

Let me describe my computer to you. *First*, I have a black box containing the *"stuff"* inside which includes over a hundred gigabytes of memory, modem, two disc drives, one of which will copy motion pictures, mother-board and things of which I shall probably never know. *Second*, I have a black 19" flat screen. *($750.00) Third* a black $100.00 remote keyboard. *Fourth* a black remote mouse on a black mouse-pad. *Fifth*, a $100.00 set of speakers complete, with black woofer under the desk. Round that off with a black *Hewlett Packard* fax machine and we have a computer-fax system done in harmonizing and inter-related, mesmerizing black. The complex sits on a natural wood desk and the units look quiet, free of wires and appear professional. Since they are in full view of my clients *(when I have any)* they look nice.

I find myself at the computer a good amount of time these days. I'm finishing Volume III of my autobiography. That is, I'm doing the final editing. I found a good way to edit written work. I separate each one of the sentences making each an individual unit and search for the right sequence of clauses and better words to express the thought. You seem to do that naturally, which I applaud, and sometimes things flow for me, too, and I have to do less editing, but to get another hard look at each sentence allows me to do the best I can do, limited by talent. I don't think I will ever be an *Annie Proulx* or a *Jhumpa Lahiri,* but there's no law against doing your best.

Briefly, because in five minutes Marge will be calling me to supper, things are going well. Everyone is happy and healthy. Marge goes to Tai Chi once a week and an exercise class three times a week. She looks great. I am still walk-running three miles, six days a week and I'm sure that slows my appointment with the ghost and the scythe. I have finished up-grading a house I did near the beach thirty years ago, building a new driveway and retaining wall and adding new wood decks with glass handrails and painting it the proper colors. The owner, contractor and I are pleased.

I'm doing upgrading on another house I did many years ago. Really! I'm going to be able to finish it! Many cabinets have to be added as well as a deck. It is an interesting job and the new owners seem glad to be working with me.

I'm also working with Kris Kristofferson, actor, songwriter and singer. You may have heard of him. He has a wonderful wife and six kids, *(grown and not-yet-grown.)* His tastes are simple and the house will be only 2,000 square feet on an acre-and-a-half overlooking the coastline. I've designed a *"what you see is what you get"* house and we're waiting for approvals from the Malibu Building Department, which, unfortunately, are not forthcoming with any speed. Malibu is in litigation with the California Coastal Commission. Jobs have been waiting in line just for Coastal Approval since last September. Approvals are not expected for another year. We wait! Kris and Lisa live in Maui in a 4,000 sq. ft. house while they wait, so I guess they are not in *too* bad shape.

Must quit. Wishing the best to you and Lucille. I'd enjoy a nice long letter revealing your remarkable thoughts.

Sincerely,
Doug

9-3-04

Dear Charlie,

Thank you for sending me the picture of yourself holding what looks like a fantastic camera. I always enjoy seeing photos of you and Lucille. It keeps me in touch with who you are and who I am talking to and how you and I are, *gulp*, aging.

Also I am delighted to have a copy of O *Enter Here to Rest an Pray*, music by J. Munson and words by C. Davis. I thought the poem highly expressive and well done. I am pleased you share your accomplishments with me. I consider them the most important duties you and I have to each other and to those who follow. They are things we have done and things we are doing with our life. I loved also your poem about your friend, teacher and mentor, *Sir Albert Brown*. How wonderful it would be to receive such a tribute from a former friend and student. I am also glad to have *The Long Day Closes*, music by Arthur Sullivan, words by my good friend, Charlie. I have Xeroxed it and given it to my computer consultant, Ron Munro, who is my colleague, musician, client, long time Presbyterian chorus singer, friend, and whose wife has been my book keeper and accountant for over 20-years. I have asked him if he can reproduce the music, either on his guitar or with piano, and sing it for me. I'd like to hear it properly performed. If *you* could play, sing it, and record it on tape, that would be even better.

Last Tuesday, August 31rst, Ron invited me to attend a dinner and lecture at the Castaway, a large restaurant and part of an exclusive hotel built half way up on the steep sides of the Foothill Mountains. *(The Foothills also form the natural backdrop for Pasadena, Altadena, and Glendale.)* The Castaway had expansive windows and decks overlooking beautiful downtown Burbank thirty miles to the business center of Los Angeles and a sweep of buildings, trees and freeways sixty miles to the Pacific. The dinner-lecture was the monthly meeting of an organization of about fifty-to-sixty members devoted to still photography called *Clicker's and Flicker's*. Ninety-four-year-old Julius Shulman was to show slides and be the guest speaker. I don't know if you are familiar with Julius, but he is internationally known for architectural photographs of the works of great architects. His clients have included works by *Frank Lloyd Wright, Richard Neutra, John Lautner, Frank*

Gherry, A. Quincy Jones, Thornton Abel, Craig Ellwood, Pierre Koenig, Killingsworth, Brady and Smith, Raphael Soriano, I. M. Pei and many others all of whom are significant names in recent architectural history.

One of the last books that displayed Shulman's architectural photographs was researched and written by a Malibu female architect friend of ours named Cory Buckner. I have the $60.00 book published in 2003 by Phaidon. It is simply called *A. Quincy Jones*. Cory gives special credit to Julius by naming three other photographers then adding, *"and especially Julius Shulman, who is a most articulate observer of postwar-era architecture and photography."*

When we arrived most of the guests were milling around ten or so tables already set. Julius, whose hair is still brown and only slightly graying at the temples, was sitting in the corner at a table near the entry selling and signing architectural books featuring his works. I decided I would like to celebrate the occasion by having one of his books. I selected a three hundred-page coffee table sized book entitled, *Julius Shulman, Architecture and its Photography*. I was surprised that it only cost $40.00. Ron bought one on Wright featuring Shulman's work for ten dollars.

Because I hoped he'd mention my name, I took my Doug Rucker name tag off my shirt and stuck it on the table in front of him so he'd know the spelling. I was astounded when he looked up and said, *"Doug Rucker?"* I acknowledged I was me and then he said, *"I know your house on the side of the hill in Malibu."* I said I was flattered he knew of my work. I reminded him that in 1965 he had shot two interiors of two houses I had designed in Santa Monica Canyon, but of course I didn't expect him to remember the shooting. He made no comment but proceeded to write *"To Doug Rucker"* and signed it *"Julius Shulman."*

Of course this pleased me greatly and was worth the trip just for this experience. We ate dinner Ron had ordered for each of us. We had salad, salmon, potatoes and green beans. While serving cheesecake for desert, Julius mounted the platform and began his presentation. He showed many 18" x 24" slides and used each to remind him of a particular episode which he would then expound upon at length. The presentation was fine for the first hour, but after almost two hours and the clock approaching

10:30 PM, the hostess, Dawn, tried several times to slow him down and remind him of the lateness of the hour. It was obvious Julius could go on forever, but he eventually caved in to prompting at 11:00 PM. Exhausted, we tipped the valet and made our way home in Ron's dark green BMW. I was astounded at Julius's energy at ninety-four and more so discovering he's *still working!*

He said Frank Gherry told him that just his three photographs of Gherry's new Disney Center Concert Hall in downtown Los Angeles told the whole story. Marge and I have been to the Concert Center with Ron and Sally. The stainless steel building is a wild, sculptural fantasy from the head of a madman. The interior is made entirely of the now depleted *Vertical Grain Douglas Fir* that, to me, has no relationship to the exterior. I liked the building as a personal sculpture, but it did not resonate with me as architecture.

What else is new? I had an infection in my elbow. Elbows are hard to see. Try to look at your elbow. I have a callus there from years of drafting and a month ago, feeling something wet, I examined it and found a small, pussy sore. I cleaned it with soap and water, doused it with alcohol, and forgot it. A few weeks later it *"blew up!"* Over two weeks later I wrote up my experience to see if I could remember what happened. I have sent this recollection to friends to explain what happened. It has been two weeks today that I have been off the drip-drug Vancomycin and as my stepson, Chris, would say, *"It's going in the right direction.*

Charlie, Marge will be returning from an exercise program she attends down the hill in a small gym. It is called *Curves* and she makes it twice around the machines that force her to push, pull, bounce, stretch, work the stomach, twist the spine and turn her body into positions it should never go. When she returns and after her shower we'll be watching news with *Peter Jennings*. So, I must quit for now. Have you been watching the Democratic and Republican conventions? I have been worrying about you two with the two hurricanes. I pray they have not hit you dead center.

Much love,
Doug

3-26-05

Dear Charlie,

   Got your fine, long letter and was pleased to hear expressed the Charlie I have known for so long. I am sorry to hear your digital camera was stolen. I am happy to have mine and that it is still working. It is a 2.5 mega pixel Olympus C-2500L, 6V. *(Probably fascinating to regular camera buffs.)* It cost a fortune many years ago, in the neighborhood of $1,500.00, but I was lucky. I got it at a marked down price about $1,100.00. Less than a year later Tom Rincker, my daughter Viveka Rucker's fiancé, got it over E-bay for six hundred dollars. Swallowing disappointment, nevertheless, I've enjoyed using it and it is not unlikely for me to take a hundred or more pictures when we are somewhere special on a single trip or hike.

   Which brings me to telling you Marge and I drove up to a growing community called Windsor, California, an hour and a half north of San Francisco. We stopped work for six days, taking two days to travel each way and staying in *Best Western Motels* for each of two nights. We visited Lilianne, my middle daughter who is married to Ernie Escovedo, a traveling teacher of occupational therapy, with a territory covering surrounding cities like Fresno, Sacramento, San Jose, Reno, Nevada, and the northwestern California territory from Oregon to San Francisco. Lili and Ernie have two boys and a girl. Cary almost 15, Nicolas just 12 and Camille who is 8. They are delightful children doing things that kids that age do. Ernie is tall, dark, and handsome with pronounced Mexican features. Lilianne is petite and as far as her appearance goes, could be the sandy-haired all-American girl. All this to tell you all the children have pronounced Mexican features and *none* looks like Lili. All look like Ernie. That doesn't stop them from being young, powerfully loved and dutifully cared for.

While there, Marge and I went for a walk in a California State Park nearby. In a little over an hour we made our way around three small lakes in the wooded California Hills. I took many pictures with my Olympus and have just viewed them on my computer screen. I set it to *slide show* and let it run. Since I usually get quite a few good shots, I will select a special one and put it on my permanent *desktop* screen-saver and select the remaining hundred or so shots on my *rotating* screen-saver. In this way, I get to experience my trip to the lakes, vicariously, all over again.

I have taken many pictures on trips and hikes, one of kite surfing. Surfers harness themselves to a large parachute-like kite sailing a few hundred feet above the ocean and shoot like the wind out to sea and along the shore and past the breakers. It's fun to watch and looks fun to do if you have an Arnold Schwartzeneger physique and the stamina of a racehorse. I've shot planes at airports, boat harbors, people and dunes at the popular Ventura Harbor, our trips to Ojai, a small town about 50 miles away that specializes in parks, art and music. I've shot birthday parties, the strange rocks at Channel Island Harbor, and many other photographic wonder-places.

I use my Olympus in my work by taking lots of pictures, mostly for my remodeling jobs. I shoot the inside, outside and for new work the site and view location photos. I *think* I'm doing a major remodeling job for Jon and Leslie Grenier in Long Beach. I say I *think*, because sometimes I've been disappointed in the past when I've been given a job and then, for one reason or another, it has folded. So far, the Grenier job looks pretty secure. Leslie Grenier grew up in a house I designed for her parents. She is the former Leslie Munro, oldest daughter of three children of Ron and Sally Munro. I designed the Munro family's house thirty years ago and then made a large second story addition over the garage and then brought it up to brand new looking just before the September 1993 brush fire when the house burned to the ground. Afterward, I designed and supervised the fire-rebuild that became one of my best houses for Ron and Sally, their three kids in the meantime having grown up were married. Leslie, who has three kids of her own from 7 to nine months seems to be a prime candidate for a client because she is convinced the of the good result she will get.

Taking pictures with my Olympus is free. I put them on my hard disk under the appropriate client or heading, and they remain there permanently. Later I can see home improvements, if any, and record them. Since my Olympus is somewhat out of date, I'm considering giving it to Marge or my youngest daughter, Amanda, and buying the new Top-of-the-line professional Olympus E300. It has 8 mega pixels and since 8 is more than 2.5, I reason the new camera will be $(8 \div 2.5) = 3.2$ times better. I like the Olympus because it has a large viewing frame and I find it easier to compose a picture if I can see it at larger scale. *Have I talked enough about cameras?*

My jobs are progressing - some painfully slow. Regarding the Kristofferson job, after the original design and two redesigns and 9-months of battling the multiple City Planning requirements of the Departments of geology, soils, health, biology, fire, public works, and road department, plus sending out notices to all neighbors within a 500-foot radius to get their OK, I finally attended the June 20th, 2005 meeting of the Malibu City Planning Commission.

It was on the consent calendar and I was pleased it passed with no discussion. Apparently, I can now begin the *hard work*. We were forced to wait until now to see if we could meet the various requirements, hoping none of the neighbors would complain or appeal the California Coastal Commissions decisions. As you may remember before continuing on the Kristofferson house we had to wait for litigation to be completed between the City and California Coastal Commission to resolve. The final result was the City is now issuing Coastal Permits according to their new rules. I've called Lisa and Kris to tell them we have passed the test and sent them an approval notice by fax. So far I've not heard back from them. *Is this going to be one of those jobs? Will they call and decide they've changed their mind? Will they be pleased with my hard work and yet, decide not to go ahead? Tune in next week for surprising answers.*

The Bright job went down the tubes. The Bright job bit the weenie. It's on the skids. It's quietly over with no shouting. Mike sold his house and his lot next door, both sites in a horrible slide zone *for 2 million dollars.* I can't believe it. I remember when people used to complain when I told them their house would cost $12.00 a square foot. They said they knew where they could get a house built for $10.00 a square foot. Mike's house was less than 2,100 square feet. His 50-foot lot next door is high in the hills with an ocean view with Palos Verdes Peninsula in the distance. It sits there with seared footings projecting from the earth from the 1993 fire. The new owners will have to sign a slide waiver if they want to build there. *(I wouldn't)* Selling his Malibu property was the best thing for Mike. It has unpleasant memories because his wife died in that house almost a year ago. He needs to get away and start a new life.

Ken Kohler has arrived on the scene. He has lived in one of my houses for 15 years. He bought it so long ago, I didn't even know he lived there. His wife was an art collector and former alcoholic. Seven years ago her depression got the best of her and while free from alcohol, yet jumped to her death from a 13th story Santa Monica window. Ken is a widower and claims no artistic sense, but has kept many of her sophisticated art pieces. We will be bringing the house up to date. Fifty-ish, he is an easygoing, athletic corporate lawyer and with an optimistic spirit. I'm enjoying working with him.

Charlie, it is Wednesday work time. I find that duty calls me. I must draw. I am sorry I didn't answer all questions and comment on your long and wonderful letter, but I want you to know I appreciate being allowed to enter your world through your letters for a short time. I am sorry about the sleep apnea. We don't get out of life alive. I can accept that, but why is it sometimes have to be so uncomfortable? Your mind is good, your writing superb and you sound healthy to me. I enjoy your letters and consider you a good and lasting friend. Thank you for adding your kind note.

No distance of place or lapse of time can lessen the friendship of those who are thoroughly persuaded of each other's worth. Robert Southey.

I second this thought. Be happy. My thoughts and love are with you.

Sincerely,
Doug

12-19-05

Dear Charlie,

Got your letter of September 6, 2005, along with the rare and probably only tape of Charlie playing the organ. I am honored with such a gift. Also, I wish to express my thanks for the booklet of Hymns called *CHARLIE PLAYS THE ORGAN*. I duly noted on the cover it says, To Doug Rucker. Since it arrived I have played the tape several times and am playing it right now while typing. You should be proud. It sounds fine. You mentioned that you still knew all of the songs you memorized. It's strange about memorizing music. That got me to thinking. I, too, know lots of songs I sang in the past from Gilbert and Sullivan's *The Gondoliers,* In Enterprise of a Martial Kind, When a Lovely Maiden Marries. From the Mikado, the Tit Willow song, and from *Trial by Jury,*

If you're anxious for to shine in the high aesthetic line as a man of culture, rare.

You must get up all the germs of transcendental terms and plant them everywhere.

You must lie among the daisies and discourse

in novel phrases of your complicated state of mind.

Oh, the meaning doesn't matter if it's only idle chatter of a transcendental kind.

And everyone will say,

As you walk your mystic way,

If this young man expresses himself in terms to deep for me, then what a very singularly deep young man this deep young man must be.

It's Secular, but still good music. I always enjoy your letters because they are candid and contemporary, right off the top of Charlie Davis's head. You don't hide your feelings. If your conversation sometimes has an edge, it's a *loveable* edge.

Things have been going at a triple-pace around here. It's an American trait I guess, though I'll have to admit in the afternoon, about 2 o'clock, I lie down and read for ten minutes before drifting off to an afternoon nap. Then when I awake, I read like a tiger for 15 minutes before heading down to my studio to do architectural work. I read a variety of things, such as two books I'm reading right now, Brian Green's, *The Fabric of the Cosmos,* and *Romo,* the autobiography of the former football linebacker, Bill Romanowski.

A few jobs that keep me going are a new house for *Kris and Lisa Kristofferson.* That one's in the Building Department. A new kitchen and remodeling job for *Darlene Beaver,* a new kitchen and refurbishing of a house for *Ken Kohler's* I designed many years ago, the replacement of a dry-rotted trellis I designed 15 years, a new garage and bedroom expansion in Long Beach for the *Greniers*, a new remodeling job for the *Johnson's* in Palos Verdes, a house expansion in Santa Barbara and four more clients I'm interviewing from calls from the enclosed publication. There are more things, but you get the picture. I'm hoping that if you know about them, you will forgive me for being so late with answering your letter. I value them you know. That said, I still have to send out 45 more cards and am doing that today. So, I better get at it. Looking forward to hearing from you and Lucille.

Merry Christmas and Happy New Year.
Doug

5-24-11

Charlie,

Thanks for the ramblin' letter. I always tell Marge, *"I just got a ramblin' letter from ol' Charlie!"* I like ramblin' letters because it gives me permission to write a ramblin' letter, too. So, here it is! I enjoyed hearing of your architectural exploits with the church. Let the person who knows what he's talking about, write. You are that person. I loved the picture you sent of you and Lucille. The view of the architectural house is great! I love it, too. I've just come up with a rule. When I design a window, I design a wall. They have to be designed together. The architect who knows the secret knows that *openings and walls are closely related.* Good show!

Regarding music and singing. During my first marriage I spent at least nine years with the *Renaissance Singers* doing motets, psalms, oratorios, masses, madrigals, chanson's and consort music. I was in a presentation of the opera *Dido and Aneas,* and a professional performance of *Rigolleto* with costumes, make-up, lighting, sets, and full orchestra. In the more secular music, my first wife and I did a month of *Gilbert and Sullivan's,* the *Gondoliers,* and enjoyed the words from *Trial by Jury*.

Later, in the Santa Monica Community Theater, we did a month of *Gershwin's, Girl Crazy*. I sang in the chorus, danced, and sang bass in the quartet.

> *I'm bidin' my time,*
> *Cause that's the kind of guy I am.*
> *While other folks grow dizzy*
> *I keep busy, bidin' my time.*

My brother, Dave, now 79, lives with his wife in Denver. He studied trumpet when he was in Austin, a 7,000 student High School we both attended in Chicago. He had his own high school band and became a football star with 11 letters for track, swimming, football, and speed skating. *(I had 7 graduating four years before.)* With a scholarship, he played football for the University of Arizona in Tucson for a year, but left because of a knee injury. Later, he played army football and trumpet in the Army's drum and bugle corps, all this, while doing a two year hitch of occupation in Germany. When released, he graduated in music from *Drake University* in De Moines, Iowa, got married, moved to Denver, had a boy and a girl in that order and switched from trumpet to stand up acoustical bass. He taught music in junior high and high schools around the Denver area for 25 or 30 years, including marching band and finally retired with a pension at 65. Since that time he's been playing stand-up bass for small groups in the area, has cut a few CD's and still remains the same old comic jock-musician he's always been.

All that to tell you I, too, came from a musical family. My dad, a tool and die maker who died in 1974 at 76, loved music. He could play the piano and sing. I can easily describe him as a *closet song and dance man*. My mother asked him how many *"tunes"* he knew. He eventually listed over 200 songs he could play and sing. I loved my dad and miss him terribly. At the piano, his right hand did all the work. It was best if you did not watch the left hand, for it wandered aimlessly back and forth, rarely striking a note. He knew all the tunes of World War 1, *Smoke Gets in Your Eyes, Sweet Kentucky Babe, Down to Getcha in a Taxi, Honey, Old Man River, Suwannee, Ezekiel Saw the Wheel* and other old time favorites. He was born just before the turn of the century in 1898,

How's that for a ramblin' letter? What's been

happening with us? Marge and I are doing quite well. I'm 83 with my 8-year-old pacemaker still keeping me alive. I'm still walking 5 days a week for over 50 minutes and feel healthy. My two best high school friends have died. Marge is on medication for Diabetes Type 2 and I guess you know all about that. For the past year she's been suffering from the beginnings of age-related tremors of the hands and arms. The diabetes exacerbates this. The tremors come and go, but it's annoying and she dreads it will interfere with her artwork in PrismaColor colored-pencil drawing. She is on a diabetes pill twice a day. The tremors come when she is either physically or mentally tense. The tremors reside when she is sleeping or calm. She still has her job one day a week as a grief counselor for *Hospice of the Conejo (Conejo means rabbit)* where she has been leading a group of about 50 whose spouses have died. She sees 15 to 20 people in each three sessions and when she retires, her patients are going to hang onto her leg begging her to stay. I'm quite proud of her work as a Therapist and also as a dancer and fine artist.

I have been doing some minor architectural work consulting and drawing for 3 or 4 former clients. One, a former client of mine in the

Malibu area, moved to the town of Potomac, next to the Potomac River and Washington DC. *(Whether George Washington tossed a silver dollar across the river at that point, I don't know. I certainly couldn't do it!)* Marge and I plan to visit the Belvin's in July coming back from a trip to Portland, Maine, where Marge's sister Bobby will be having a family reunion for her 80th birthday party.

Another client is Bob Ross. I completed a master bedroom addition to his house that overlooks 2 miles of white-water shore break along the famous Zuma Beach. He can also see 180 degrees of Pacific Ocean, Point Dume, a promontory that marks the western end of Santa Monica Bay, and Islands, Catalina Island, Santa Barbara, Anacapa, Santa Cruse, and San Nicolas. The addition has an 18-foot glass door that opens completely in the jack-knife mode. My present task is to do the landscaping around an existing pool and spa. I love doing it. It's a rewarding job.

Also, I just completed stage 3 of a renovation of a 1,600 sq. ft. house I designed about 40-years ago. Almost completed, this house is fast becoming what it was supposed to be years ago.

This time I have a more sophisticated owner.

The fourth job is really slow. It's an addition of a large storage room over an existing garage. I've been working with this client for over 30-years and his beach house, after all this time, is really in good shape. We are waiting interminably for a Planning Approval from the Malibu Planning Department. I'm not holding my breath, but it will eventually happen.

In between my consulting, and believe me, there's a lot of in-between, I am devoting my time to doing artwork in Digital Photography and trying to get my books published without spending a fortune. Jack Birdsall is a friend who loves to publish books and has been working with me to try to get them published. I have one book finished and have numerous books ready to go. They're in the computer, but not properly formulated. I will send you *Book of Words*. I think you'll like it. Read it with tongue in cheek.

My new book, 80% completed, is a philosophic book called *Harold and the Acid Sea of Reality*. In another comedy book called *Further Adventures*, I start off with 8-pages of cartoons where a person diligently follows the signs he's supposed to follow that lead him into a brick wall to which he dutifully bumps into. Half way through the book is another 8-page cartoon with views of outer space. As we get closer, we recognize Santa Monica Bay, and closer yet to a guy that holds up a bank. As we recede, we wind up again in outer space where a view of the earth is lost among all the stars. *(Moral: If you're on the earth, watch what you do.)* At the end, another 8-page cartoon, where a guy with a briefcase walks along without noticing a huge sign that points irrevocably in the special direction he is going. He misses turn after turn until a lady rescues him by pulling him over to her picnic blanket, while the huge sign turns into a gorgeous tree.

Digital photography is an everlasting journey of discovery for me. I take pictures of plate glass windows in commercial areas and malls. What I get is not only what's behind the glass, but the reflections in the glass of trees, buildings, cars, trucks, bicycles, accidental passersby, etc. Since my 600-dollar camera with 12.1 million pixels will make a clear picture about 5-feet square, I can frequently take a portion one or two feet square that is abstract or filled with exciting variations and make a picture out of it.

In the past year, I've been in well over a dozen art shows in galleries in Ventura, Camarillo, Westlake Village, Thousand Oaks and Ojai. I've just been invited by the Artist's Union Gallery, of which I'm a member, to give a 15 - 30 minute talk on my present artistic approach. The official thinks I've discovered something new. Though I'm not much of a speaker, I'm flattered and the ego part of it will probably get to me.

I enjoyed, immensely, your ramblin' letter. Don't ever change. Am running out of ramblin', too, and have also enjoyed this visit. It was wonderful to hear from you. Please write soon again.

Sincerely,
Doug

3-18-13

Dear Charlie,

    I was delighted to get your letter. Your first paragraph was almost a complete letter by itself. *First*, a lesson on the spelling of Mahican's as against Mohican's, but my *"spell-check"* allows Mohican's, so maybe either spelling is OK. *Second*, you've reached the age of 94 without losing any of your verbosity. I'm just a whippersnapper being nine years younger at 85. I'm glad to be old - especially considering the alternatives. *Third*, just as always, by writing me a letter you can still let me know how you feel and let the devil take the hindmost! By the way, that's a compliment. I appreciate your obstinacy and as well because I can be the same way. *Fourth*, I love that you follow your own rules and your own heart. I've tried to teach my kids to do the same, and I must say, they're doing pretty well. *Fifth*, as all men do, you follow the dictates of your stomach and listen carefully for any sign that announces dinner. Lucille is not only *special*, she's *vital*. *Sixth,* Your special hobby, playing the organ. Music! Music! Music! How can I love it so? How can I live without it? Yes! You said all this in a simple, 10-line, first paragraph. It's enough letters to answer all by itself.

Your *Claude Monet Impressionist* and probable plenaries landscape with which you cover yourself is a *classic*. At our age, why not? Claude died in 1926 just a year before I was born. I, too, am familiar with age-related difficulties. Yours: incontinence, heart attack, sleep apnea and diabetes. Mine: stroke, pacemaker, elbow infection maybe flesh-eating disease, high blood pressure, broken rib with punctured a lung, ad-infinitum. I, too, take a plethora of pills. Mostly for cholesterol and high blood pressure, but I also take supplements like zinc, saw palmetto, and glucosomine-condroiton. I still walk 45 minutes a day for 5 days a week, but it's actually more like 3 days a week. My hips hurt, but I walk through pain. Also, when I was having emotional problems following my first marriage, I thought it compulsory to internalize Omar Khayyam's poem, *The moving finger writes, and having writ, moves on, etc. - A wreck at last must mark the end of each and all.* Though it's comforting to think that when we die, we've lived our *entire life*. I've been thinking a lot about death in the last few years and have just finished a book called *Proof of Heaven*, which was supposedly going to give me hope, though I still have concerns. I just finished a Margaret Isherwood 1964 book I felt well worth reading called *Faith without Dogma*, who, by the way, is still living in the South at 94.

I'm glad you like your new Pastor and do hope you will add new members to the church. I thoroughly enjoyed your poem entitled *HALF-WAY PLACE*. It's an artwork with rhythm, rhyme, recall, and meaning. I like the idea; I only made it halfway, but it's not a curse, you'll fill it with life while it still is yours. Many years ago I finished writing a 467-page book of poetry called *MOVING THROUGH*. Your *HALF-WAY PLACE* poem reminded me of two poems I wrote in 1967 and 1976. An illustration follows:

MORE NOTHING - 2/77

I shall proclaim it to the hills,
sing it to the flowers in the fields,
shout it to the multitudes
that gather in the streets.
It shall be heard,
my nothing.

For I will have it heard.
I shall cry it to the gull
that hovers on the wind,
and to the stalwart trees.

(I'll bring them to their knees.)
I'll sing my message softly,
to the violets in bloom
that arch their dainty heads
to escape the forest gloom.
I'll whisper it to frogs and snails.
(Tie a message to their tails.)
They'll know the tale
that I must tell,
my nothing.

And when the winter winds
come whistling through the fields,
come crying through the backyards
and open country fields,
and then the sloping snowdrifts
and crisp and crystal night
comes settling on the surface
of some special starlit night,
I'll stand alone and think,
project my message strong and clear
so all the universe will hear,
(to me it's very dear)
my nothing.

STRANGE - 4/67

A mouth and nose
to taste and smell
and ears to hear the sound.
Skin and fingers,
sense of touch,
and eyes to see around.

So, Charlie, I think we're aware of life. It's good you've compiled a loose-leaf book of your writings and took the trouble to share them with an old *Last Mohican* friend from the *University of Illinois*. You have proved to me by your letter, poetry and our correspondence that you're an excellent writer. At our age, we get to say anything we want, right?

I enjoyed the coverage of your schooling and early life: stories of your work building airfields in Europe and tales of your marriage, family and early office. You said you think you made a mistake setting up a one man architectural office in Roanoke. I presume that is because of the territory's conservatism. Your stories of your work with the National Park Service and in Fort Bragg and erecting telephone poles, bridges and concrete bunkers shows a good bit of living having taken place. I very much enjoyed your narrative of that time in your life. I'd forgotten you swam competitively on your team in High School. In 1945 I won thirds in the Chicago High School City meet for the 50 and 100-yard sprints and we won both 4-man and 3-man relays with me swimming anchor. As you know, I then swam for two years for Illinois, and in any case, enjoyed reading of your life in only six pages. You can fill me in on details later.

As for other stories about what's going on around here, first, Marge at 79 has her diabetes fairly under control. The Doctor has recently taken her off pills and she just treats herself with diet and exercise. With a complaint of swollen lymph nodes in her groin this last year, she was diagnosed with a lymphoma and has just finished her 5th weekly infusion of Rituxan - a drug that only fights the bad guys and leaves the good guys alone. She feels pretty good after the infusions and isn't nauseous and won't lose her hair. What's almost as hard or, perhaps, even harder, is the fact that she has the beginnings of Parkinson's disease. Her hands want to shake all the time, an especially difficult problem for an accomplished PrismaColor Pencil artist. When she's awake, they shake, but when she's sleeping, they stop. When she tries to relax she can sometimes stop the shaking. Hands shaking are difficult, but she's stoic. If you met her right now, I doubt you'd know she was having physical problems. Despite her long naps in the afternoon, she pulls off social Gallery openings quiet well. She's now retired from being a Grief Counselor for 27-years at the Thousand Oaks Hospice of the Conejo. Conejo means rabbit in Spanish - I don't know how that came about. When she's not doing Doctor stuff, she works on taxes, bookkeeping, pays bills, goes shopping, does cooking and is a marvelous companion. But the combination of putting up with symptoms from all three rather severe illnesses including just age at 79, take their toll. Nevertheless, I continue to be madly in love with her.

After being idle for three years during the countries financial downturn, I've bounced out of retirement and have a small architectural job or two. The old bell went off and I dashed into the ring and started swingin'. It seems my past has come back to haunt me. When one or two or three of my houses came up for sale, Real Estate Agents are advertising them as another Doug Rucker Malibu Classic. Apparently, I have done *"Classic"* houses. One job that I am adding to is an addition to a house I did 53 years ago. It still looks pretty good. The owner built it in 1960 for 30 thousand dollars and the owners just bought it for a million-three. Where does an alien go to register? The other job is trying to make a new house out of a 2400 sq. ft. barn-like structure. Isn't *that* fun?

Also, I'm continuing showing my reflective photography artwork and have three pieces in the *Ventura County Atrium Gallery* and five pieces in the *Ventura Community Memorial Hospital.* In three years I have shown over fifty times in Ojai, Ventura, Camarillo, Westlake Village, Oxnard and Thousand Oaks. Over that period, Marge has shown, too, and recently garnered two first prizes.

Also, I've finished three new books, *Reflections, Book of Words, and Where's the Cookie's At?* Two new books are in the slot. An old one called *Personal Journey* - a poetry/prose book about my divorce in 1982 and a new book called *Harold and the Acid Sea of Reality*, a poetry/prose book on the difference between fantasy and reality. The books are published by LULU, an *"on demand"* publishing company, where, if you send them a PDF manuscript and cover designs by Email, they'll print a 6in x 9in, 250-page book for about $12.50. On my computer under GOOGLE, when I type in LULU/Doug Rucker, I get *Early Stories*, a book I think you have, and clicking that, a list of five other books. Just below, you should see a listing for an architectural house of mine I did in the 60's. If you clicked this listing, you would see the old *"classic"* house with about 30 additional pictures.

Well, I guess I'm winding down and I *do* have other things that are pressing. My kids seem to be doing well. Viveka, my oldest at 53, is married and is a Nationally licensed Acupuncturist doing business with a partner in Woodland Hills, California. Amanda, my youngest at 49, is a single mom raising two teen-agers, while being a registered Nurse doing home health care and working in the Malibu area. Lilliane, my middle girl who is 51 has three children, one college age, one in Junior College and one a senior in High School. She is a Practitioner for the Center for Spiritual Living, a non-denominational church in Windsor. Windsor is a small town about a two-hour drive North of San Francisco. Marge, of course, is a *Marriage Family Child Counselor* and so you see, I'm well taken care of in body and soul. I'm proud of my kids and pleased they're still talking to me. Marge and I will be going to Kauai for Kainani's High School graduation this May and that's the news from Lake Woebegone.

Enjoyed your letter, Charlie, and will look forward to your viewpoint, comments, freedom of your one-of-a-kind thoughts, and any wild thing you want to get off your chest. This note is from one Mohican to another with love to Lucille. By the way, has your eyesight improved?

Sincerely,
Doug

CHUCK HAZLEWOOD FORWARD

At 22 years old I got my first job after graduating from the *University of Illinois, and* I finally worked in Denver, Colorado with an elderly architect named E. G. Groves. Also working for Groves was Chuck Hazlewood, a 25-year-old good looking, blond, athletic, draftsman. He befriended me and patiently answered my architectural questions. He had been in the Naval Air Corps, was married with one child and as Chuck admits, *"admired feminine pulchritude."* If this could be considered a fault, I ignored it. We became good friends. Also in Groves office was a 30-year-old married person named Don Wiederspan who'd been in the *U.S. Coast Guard*. Also, Paul Graves, a thin person with glasses and former *U. S. Army* man I came to know as Grove's long-time right hand man. He was the principal in charge of the office when Groves was out.

Though Groves office was not without benefit, it was not my first choice as a place to work. In Groves office I remember doing an ink tracing of a floor plan of a rather big, clunky looking building. After that, I did whatever I was required to do even though I don't remember what. Some people say you mentally block what you don't like. The architectural work was unsatisfying to me in every way, but I needed office experience and in 1950 this seemed the only job available. It's no wonder I blocked what I was doing there.

Chuck Hazlewood soon became someone I could like and trust. He complimented me when he came to know me better, saying I had many good qualities, but needed more experience with girls. He called me *a social diamond in the rough.*

During our lunch hour we'd walk along the busy downtown Denver streets packed with shoppers, young and old and other white-collar workers out to lunch. While walking and talking, I was sometimes surprised to find I was speaking to air when I'd notice Chuck had quickened his pace moving ahead to get a better look at some beautiful young lady.

Despite his admiration of *"pulchritude,"* he was careful to be discrete and was at all times a gentleman.

Nancy, Chuck and Mary Ann Hazlewood

Hazlewood and Wiederspan introduced me to a rather sophisticated, somewhat taller, young blond girl who worked nearby. She was quite beautiful and available. I lacked the courage to ask her for a date, which drove the two married men up the walls.

"How can you be so stupid? They asked. There she is, literally asking you to go out with her and you won't give her a tumble. You're crazy!" They'd moan and roll their eyes and slap their foreheads. Chuck was the most sensitive to my feelings, but didn't choose to embarrass me further. He knew I was concerned about my shyness and that it would take me some time to improve.

In good time Hazlewood and I started playing early morning golf. First we went to the driving range and I learned to hit a few balls. After several sessions, we agreed to try nine holes. I can remember awakening at 5:00 AM and getting to the course by 6:00 AM and playing until 7:00 AM, then having breakfast and getting to work by 8:00 AM. Though I could belt the little white ball quite far for my size and age, nevertheless, my scores were in the middle 50's for 9 holes.

After eight months working in an un-satisfying Groves office and having finally acquired a car, I decided to try my luck elsewhere and head south to Tuscon and eventually to San Diego for undiscovered pursuits.

2-17-99

Dear Chuck,

   I was so taken by your letter, particularly the dedication that *I had left footprints on your heart.* I almost broke up. That was a very nice thing to say. You can't buy that kind of praise. I had to write you to tell you how much I appreciate that fine compliment and the wonderful feedback on my book, *Early Stories.* Also, I'm glad you could find things in the book with which to relate. I'm continuing the book because I wanted to see for myself what I'd done with my life. If you can find a reader, the next pleasure is to find out that what happened in my life reminded the reader what happened in his life. Then a dialogue will be started.

   Of course, since we spent time together, we can remember our viewpoint of that mutual experience and the book is more ours than mine. I've already written about my experiences in Denver. Chuck Hazlewood is very prominent. I've even included the last letter you wrote in response to my 70th birthday. That is also why the picture you sent me of yourself, Mary Ann and Nancy is so important to me. It is perfectly safe for the time being and unless it's urgent, I would prefer to reproduce it when I reproduce my other photographs. Let me know if you agree.

But, let me fill you in on what's taking place. I now have architectural work. I am about 3/4 through the working drawings on a house fire re-build. The Owner's, whose site is on a beautiful promontory a mile as the crow flies off the ocean and a thousand feet up, lost the roof and interiors of their house made of concrete block in the brush fire of 1996. After a couple of years dealing with the insurance companies, they contacted me to do the architectural work for rebuilding. Only the concrete block walls, slab and foundation still existed. I was contacted because I had done a 450 sq. ft. addition to the house about 15 years ago for the wife's parents. Her parents died of old age and she and her husband inherited and occupied the house. They saw my name on the addition drawings and since *"it was so beautifully done,"* how could they not call me.

I am also doing a landscape plan for another house I had designed in 1970. It's about time they got to the landscaping. I am also helping a former client with some water damage repair on his house on Malibu Road. I also just finished preparing as-built drawings and writing a program for a female member of the *Malibu City Council*. We are on hold for an undetermined time on that one. Also, I'm helping Viveka's fiancée, Tom, by doing design and drawings for moving his sound studio out of his house and into his garage.

This is why I have made no progress since December on Volume II of my autobiography. I enjoy the architectural work and will continue to work as long as I get jobs. Running a close second is Autobiography Volume II.

Volume II and Volume III and Volume IV, if necessary, will be a continuation of my personal life and that which my kids will, I hope, enjoy. It will be a kind of history of, myself in architecture. I'll be seeing how I came into residential architecture, how I obtained the vital philosophies, what mistakes were inevitable given my personal make-up and what it takes to become a semi-successful architect. In those sections will be photographs of my work and interesting stories about many of my clients. I'm looking forward to it and I'm very happy you are interested to continue with me.

I appreciate your fine words and when Marge and I are in Golden, we will call you. Chuck, I can tell by your letters you are a good writer. I would appreciate hearing some of your life stories. After all, mine are just my *take* on my life. On your life, what is your *take?*

Best wishes,
Doug

7-4-00

Dear Chuck,

   Well, here I am returning your picture about eighteen months after I said I would. But then I'm hoping you are healthy as ever and have not been too bothered by life as it's lived. Please thank your daughter, whom I presume is Nancy Hollaine, for digging it out. It is just what I needed and I've included it in the enclosed a first autobiographical chapter. If there are errors or omissions you see, I'd appreciate it if you'd let me know so I can correct them. The writing in volume II is just about done and I am inserting the pictures in order to see where and how it will end.

   While returning from a visit to San Diego to see Marge's sister, Barbara, Marge and I were trying to think of a title for the second autobiographical book. We think we've got one — *Groundwork*. Hot! Eh? In Volume II, I not only do groundwork for my life, like getting married and setting up my practice, but that's also the first thing you do when you build something. And since we are both in the business of building something — well, you see . . .

The last time I wrote I spoke of doing a house on a promontory. It's finishing up nicely and the Owner still talks to me, so I guess we're doing all right. The interior is dry-walled and the glass is in. Next week I hope the cabinets and roof will be installed. It's running about $420,000.00.

Also I have an addition to a house now building. It's a $60,000.00 job with a dining room extension and new kitchen with breakfast bar. I did the original house about 25 years ago, a post and beam for some people called Terrill. They sold it after 5 or 6 years to the Wolf's. Then I did a thousand square foot *(2 bedrooms, bath and work room)* for them about 10 years ago. This is the third time I've worked on this house.

Malibu has a laundry list of *May-I's* to go through: Geology, Soils, Private Sewage Disposal, Coastal Commission Fire Department, Highway Department, Biology Division, Archeology Division and on and on. It's not like the good old days. I'm trying to get permission to build a new house for clients of mine who've decided to move out of a $700,000.00 house I designed for them 22 years ago.

It's interesting that is what they choose to do

with their lives. So, I'm in that process, as well.

Then some friends I've known for 28 years have decided to upgrade their house, *($100,000.00)* and so I'm to do the remodeling. This architectural activity, I'm both happy and sorry to say, kept me from working on my book. But today the July 3rd and tomorrow, the 4th of July, I'm making progress. For the book, I will be photographing a photograph of me in my first office. It is a profile silhouetted against floor to ceiling glass overlooking the ocean. I'm apparently waiting for work.

As for family news - Viveka, my eldest daughter, is living with a wonderful guy named Tom Rincker, while she continues her studies to be a doctor of *Oriental and American Medicine.* She will be 40 this September and she and Tom plan marriage sometime after her graduation.
Lili, my middle daughter, is raising her three children, Cary *(10)*, Nicolas *(6)*, and Camille *(3)*. You'd think that was all she could handle, but she is working part time as well. She and her husband, Ernie, an *Occupational Therapist,* live in Petaluma, 40 miles North of San Francisco. Amanda, my youngest who will be 36 this September, is married to Scott Jensen and they

expect their first baby *(a boy)* in September. So, that's the family news — except for Marge's four kids.

I'll enclose an article about Marge, which tells something of what she's doing. In addition to her job she sees a couple of clients in her greenhouse studio and recently has been getting back into artwork. She's doing PrismaColor drawings of meaningful objects in her life, like a cream pitcher, or a unicorn, or a house, or the underground view of the New York subway system, or the Rockefeller Center sculpture and putting the finished works in a 20"x30" portfolio. Like my writing, she considers it a life review.

Marge is calling me for dinner. It's been nice chatting with you. Give my best to Bernice. Thank you again and thank your daughter for the picture. Next letter, tell me what's happening with Hazlewood.

Sincerely,
Doug

8-26-00

Dear Chuck,

Sorry it has taken me so long to answer your letters. I have been busy with my work and haven't found time to do any work on either my book or your letter.

First of all, you should know you were an important figure in helping me resolve one of my main conflicts — a correct association with women. As it was, I married someone much like my mother and finally after 25-years I was divorced. As you can see, you weren't the complete cure. You were a person to whom I could relate, who never belittled me or made fun of me, but accepted me, my faults, and my better parts as a complete person. I'll never forget that because that's the way it was in 1950.

Of course in fifty years we each can change. I consider you to be one of my better friends in my whole life. To do anything leading to a destruction of that feeling is something I'd like to avoid. My memoirs are my memoirs. If they are in error, it is more important for them to be truthful. That said I have willingly changed some of them so they will more accurately align with how you saw yourself and to admit, perhaps, I saw things inaccurately.

Please be mindful that this book, *Groundwork*, my second autobiographical work, would amount to 40-copies sent to Marge's and my kids, a few cousins I rarely see, my brother and a few friends and architectural buddies who have worked for me in the past.

I plan to open the book with a preface that will contain something like the following:

It is my belief we create memories to suit our survival and comfort needs at the time of remembering. That we rarely recall the truth on what took place, but based on those events, we rearrange, reorganize, embellish and change our recollections in the way we need or want them to be. These new remembrances and images then become the reality forming the basis of our new actions. What follows (the book) is my truth and I can readily accept the idea that it wasn't the same truth to another. Please forgive me if I have left anything out or have not seen or expressed what might be another truth.

For instance, I had forgotten until you wrote

me, that Groves was a member of the *Church of Divine Science*. I remembered him as a *Mind Scientist*, whatever that means. Thanks for reminding me. I remember him believing whatever he believed would occur that's radical in my mind. I am an agnostic. I had just finished reading both the new and old testaments of the bible and found, therein, many contradictions. I thought the Bible shouldn't be taken literally. So you might say I was against most parts of The Christian organized religion.

Groves belief that I wouldn't leave his office produced an *"Oh Yeah?,* attitude in me and an *"I'll show you!"* response. I was flattered he wanted me to stay, although it was a mystery why he wanted me to stay. I was affronted with his audacity that he could control me. *(Yes! I was young.)* Even so, I gave him his due as I saw it. See pages 26 and 27, enclosed. Thanks for reminding me about his invention of the prefab concrete house. I'd forgotten.

Radio Psychologist, David Viscott, whom I admired and who died of diabetes a few years back, defines *belief* as *"knowing something is so whether or not it is."* And he defines *knowing* as *"the truth from a certain perspective."*

My autobiography is a truth from a certain perspective. If you remind me of something and I recall that, it becomes my truth — therefore, so are my revisions.

There were many good things to be derived from Grove's office: detailing and specifications, office management, organization or, considering Grove's desk, disorganization, being in the business, etc. In writing my book I didn't mean all who worked for Groves were like Groves. We each have our priorities. I was young and carefree. *(Carefree?)* You had Mary Ann and Nancy, your new child and these were important considerations. At that time I was a lost kid in a man's world. I was a person not looking for work, but looking for dating and some kind of social maturity. The things we wanted out of life were different. I'd be interested in a written testimony to your own life. What are your memoirs? How would you describe Grove's office? What were your concerns during that period of your life? You write excellently and have had a diversity of experiences. Your book successes, problems and failures would be a good one. I'd read it.

I think my experience in Grove's office will be clarified when in context with the next few chapters. Groves and Chuck were necessary persons to me - each on my path.

Chuck, you were a key figure in my life, helping kick-off *(start)* my social development. However much I achieved, it hinged on your sensitive and substantially kind attitude toward me. You modeled what I needed most at that time - that women were OK and that you had a general love for all women. I will always be grateful for the relationship we had *(and have)* and give you credit as my first teacher on how to behave as a mature adult in what then seemed to me a foreign world. Without you in my memoirs, I would be missing a special and important person. *(Sorry, you're doomed!)* In fact, with your permission, I might like to put additional facts about Groves as an addenda at the end of the book: the condition of his desk and Bob Harley, the Oldsmobile, 0-to-60-in-20-seconds and likeness to Barney Oldfield, thumbnail sketches on the backs of envelopes, 7-story building in 7 different classic styles, process he called *Concreter with Doors* that closed like cushions.

Thanks for the information that all who worked for Groves were in the Armed Services and all glad for my deferment. I've changed *fled* to *flew*.

On the personal scene, I was sorry to hear of your break-up with Bernice. It sounds like you're recuperating well, especially from Smitty who, among other things, you've taught to love flying. Chuck, I didn't know you were a flyer. How long have you been flying? How many hours do you have? What kinds of planes did you fly? I didn't know you did television commercials. Who were the advertisers? Do you get to work with any starlets? Did you like it? Is there money in commercials?

I keep busy. I'm supervising a fire re-build in Malibu and an addition in a nearby community called *Montenido*. I've just finished preliminaries on a kitchen, dining, master bedroom with two baths and a 2-outdoor patio addition. We'll be getting preliminary estimates, *(guesses)* before we proceed.

I'm still jogging three miles, three or four days a week. For the past year I've been dealing with acupuncture once every two weeks, daily herbs and Tai Chi once a week as a prescription for an irregular and slow heartbeat. I have been noticing a benefit so I'm staying with the program.

My oldest daughter, Viveka, who is 40 this

September, is three years into becoming a *Doctor of Acupuncture* combining American and Oriental medicine. I'm very proud of her. My youngest daughter, Amanda, *(36)* is due for her first baby boy early in September. We've already given her a shower. My middle daughter, Lilianne, is busy raising her two boys, 7 and 10, and daughter, Camille 3, with her husband, Ernie, in Peteluma, near San Francisco.

Say hello to Wiederspan and let me know if I'm ready to print. It doesn't print until *you* say it prints.

Sincerely,
Doug

10-17-00

Dear Chuck,

Your letter was gratefully received. I am late in writing because of my naiveté about Gmail. Computers are known to be so fast that I expected e-mail letters to be on the screen instantly when I signed on to mailbox. I'd sign on, call up mailbox, see a blank page, and go to my next project without noticing the little box below telling me the computer was dutifully printing Gmail. This morning I am beginning to answer 12 *(count 'em — 12)* friends, contractors, kids, and even one grandchild, all of whom were trying to contact me. Yours is the first and I owe you an apology.

My Coral Canyon house is about finished and thank the lord, after two and a half years, the owners are pleased with the work. My *Montenido* addition is also about finished and I might be getting a new job this week. If so, it will be a game room addition to an existing contemporary hillside house. It is in an area where I have completed five houses and two remodeling projects over the last 25 years. Three of the houses had been completely destroyed in the November 1993 brushfire, but the owner's were able to do much better houses for themselves the second time. They had more money, were no longer raising a family and I had more experience. I'd like you to see them next time you're in California.

Thanks for the background on your flying experiences and your acting. I was blown away by the fact that you were once a stand-in for Jack Warden. Jack was a client of mine in 1960, just after I opened my first office in Malibu. A family from Phoenix had asked me to do a beach house in the *"Malibu Movie Colony." (So named because that's where all the silent film stars and director's used to build their get-away beach homes.)* Upon finishing the Phoenix family's house, the owners of the house next door, Jack Warden and his then new spouse, the gorgeous centerfold type, Vonda, asked me to do a guest room over their existing garage. I thought I did a good job and Jack seemed pleased with it during the little time he was around.

I always liked Jack. He was very friendly to me, even though I found it hard to get to know him. During construction I'd walk into his house for the answers to questions or to give him a personal progress report and he'd offer me a glass of beer. I'd say no and he'd offer a coke. I'd say OK. Then he'd ask if I wanted a sandwich. If it was lunch time, I'd say Ok and we'd sit on the couch in front of his fireplace *(which I never saw lighted)* and he'd say, *"How's it goin'?"* I'd say, *"Fine."* Then he'd say, *"Yep!"* and eat a bite of his sandwich. And so it went through the lunch without anything to talk about. Soon I was on my way.

I have corrected the *"man"* and changed Mind Scientist to *Church of Divine Science*, corrected the spelling of Don's last name to Wiederspan, and corrected the duplicated picture on page 21 and 22. At this time it is my intention to finish the book and pictures for a proof copy, then re-read it carefully and have Marge and probably my oldest daughter, Viveka, read it again for corrections, typos and foolish mistakes. The copy you have is sort of a *"proof"* copy. I do appreciate you reading my work with a careful eye and correcting my spelling and typographical *(and other)* errors. It's never too early. You should know that I printed only 40 copies of *Early Stories*. This time I may print 50 copies. The book is not slated for general publication and I'm sure will get no reviews and *"0"* notice. I send the book to my reading friends and living relatives whether they're curious or not. At least a quarter of them failed to acknowledge they even got the book. What does that tell me? I guess reading something about someone else is a drag.

I do have one or two fans, though. Several have told me they enjoyed the book, thought it was unique and were looking forward to the next volume. I admit I have been pleased to hear this. I realize most will not tell me if they find fault. Actually, I'm not doing the book for them. That is not to say I do not care about the people I write about. I do care intensely and think I have taken pains not to hurt any of my friend's feelings. That is *not* the purpose of these books. In writing the stories of my life, I just want to remember things that happened to me and to see if that fresh look will change my approach to how I will live the next part of my life - if I have a next part. *(I had a limited stroke in 1993 and have recently had marginal experiences with bradychardia that I am now treating.)* I know my life was worth it and I have never, nor am I now taking it for granted. Still, I want to question my life in relationship to what it yet might be. If it needs a change, I want to change it. The books are for studying and remembering where I've been. What to do in the next period of my life comes next. Like it, or not, you were a positive person at a critical transition in my life.

Let's stay in touch. Marge and I are not planning a trip to Denver in the near future, but we both would enjoy having you and Smitty visit us when you're in Sunny Cal. Just think, you'd get to sleep in our master bedroom queen-sized bunk bed by the window. When we have guests, Marge and I vacation in the comfortable bed in our greenhouse.

Sincerely and appreciatively,
Doug Rucker

9-12-01

Dear Chuck,

   The longer, more literary, and more-smarter letter of merit, counting at least double over the former more-cursory scribble of thanks, is now upon you. My grammar monitor just turned the previous *more-smarter* letters *green*. What does that mean?

   That brings me to *Microsoft Word.* Both books were done in the *Microsoft Word* program. Now I have learned that it should have been *typed and edited* in Microsoft and *transferred* to *QuarkXPress or PageMaker*. Dumb me! Book three will be installed in QuarkXPress before printing. In Microsoft it took me months to edit my own work. When I finished the Master Copy, I assumed everything would print accordingly, but when I printed 45 copies, I discovered new errors not in the Master Copy. Microsoft, I have learned, is not an efficient program in which to write an entire book with 150 photographs.

   I am including with this letter an e-mail copy of a response I wrote you regarding your letter of September 5, 2000. If you didn't receive my earlier e-mail response, I probably didn't send it correctly. Proper use of Gmail or Internet is not one of my highest priorities. I know I'm missing a lot, but then I have other things to occupy my time, like architecture, five current books, seven children and their spouses in one state of marriage or another, seven grandchildren, and a relationship and house to keep up. They come first. Anyway, I like to receive and send letters by mail.

Regarding your letter. You said I seemed dedicated to architecture, so much so, I had to *immediately* leave Grove's office. I have to admit — *and this I'm finding through writing an autobiography* — my principle agenda at that time was growing up socially, not pursuing architecture. During the Groves period I was still giving myself permission to leave the whole architectural field. It hadn't occurred to me that parts of me were well suited to the architectural profession.

It was interesting for me to learn you'd been with Muchow for seven years. *(By the way, when, how, where and what age did Groves die? He did die or didn't he? Is that why you left? I'm curious as to the rest of the Groves story?)* With Muchow, you must have done some exciting projects.

Wasn't Muchow doing mostly commercial work? It is interesting to learn he had a small office rather than one, as I'd imagined, of ten or so draftsmen.

I did many strange things in that 22 year old adolescent period of my life. In Denver, I did not contact Muchow. I didn't know it then, but at that time in my life I was more in search of girls and adventure than searching to become an architect. Architecture seemed the only way to pay the way. First I found out what Denver was like and missed Muchow. Then I tried Tucson and found an Illinois alumnus willing to hire me. I refused. Tucson was too hot. I wanted to see the ocean. Led by the old unconscious, struggling by a method Marge calls *Uga-Buga,* I blundered to San Diego. Later in life a writer was interviewing me for an entire one-page *Malibu Times* newspaper article about my works and me. She looked at me in what I interpreted to be awe and said, *"Mr. Rucker, when did you first know you wanted to become an architect?"* I thought carefully for a moment, then replied, *"When I was about fifty."* She assumed I'd always wanted to be an architect. Not true. I got into architecture because my mother thought I'd do well in it. Final decision, blame *Mom!*

Regarding disillusionment and maturity. I feel I'm an expert on it. I'm going to make a feature out of it in my next three books - if I live that long. Talking about finding one's self: I know one way to get a good look at my real self. If I can say about anything *"I'll never do that again!"* I get a momentary glimpse of my real self, whether I like it or not. Too bad if that's the only way I have to find it.

Regarding God and understanding the cosmic intelligence of friends and clients, I am a deeply religious and spiritual person and believe in all religions, but not one specifically. I'm attuned more toward Buddhism than anything else, but I have not studied it with any enthusiasm. I don't generally like religious organizations. To me, they are slowed by their own rules and find it difficult to change with the vastly changing world. In the book, the word *God* to me is easier to understand than *Great Life Force*, which to me is the same. When I say God, you may substitute if you wish *Great Life Force*.

What I infer from your asking how I knew everyone's height is that, either I have a good memory, or a person's height gives a better description of the person, or in so expressing myself in terms of how high people are, I show a personal inferiority to taller persons. I don't know about memory or accurate descriptions, but I don't think I feel inferior to those taller. *(Tom Selleck the exception.)* In football the motto is *the bigger they come, the harder they fall.* Our high school football team was outweighed ten pounds in the backfield and twenty pounds in the line by our leading competitor, yet we still tied for the City Championship in *Soldier's Field* in Chicago. I have a beautiful wife who is five-feet-three inches tall, perfect for me. Also don't forget I have the big disclaimer in the *Forward* to *Groundwork*. By the way, in 1950 I was five-feet seven and one-quarter inches tall and weighed a hundred and sixty pounds. Now I'm five feet six and a quarter and weigh one hundred forty-eight and a half pounds. *I get older. I get smaller. Soon I'll be gone.* I have several trees I've planted on my property and though they are automatically watered, the soil is so bad they get shorter each year. I figure I'll get four or five years out of them before they disappear completely.

I did complete the project where I had to compete against Quincy Jones. The pilot house was built and I'm including a couple of photos to show you what it looks like thirty-five years later. Too bad it is covered with trees and foliage and the wood painted brown rather than left in its natural color. The house was quite handsome with a high, open-beamed, gabled roof. It had a huge glulam ridge beam with a see-through sculpture hanging from the ridge guiding water to an interior pool and garden below. The pool, garden and sculpture was designed by landscape architect, Eric Armstrong, a business associate and friend who did major landscape work at *UCLA* and *Santa Barbara College* for years. *(See pages 208 through 210, Groundwork.)* When passing from room to room in the house, the spilling waterfall was intended to be an architectural delight. At least I thought it was. Either that, or the owner's hated it. I don't know, I never talked to any of the occupants. The waterfall did not occur. That story will be in the next book.

Regarding the office you visited, yes, it was right on the beach in the *Tide Pool Gallery* building that was then called the *Chris-Craft Building*. Can you believe the designer was the young John Lautner? The *Chris-Craft* building was abandoned unfinished. Someone never put on the finish coat of plaster and I often wondered if John Lautner was sufficiently paid for his designs. My office was in two sections of that building at different times. Both were on the second floor. You must have seen my first location. *(Page 189 to 190, Groundwork.)* The second location was also on the second floor and was directly over the water. *(Page 210 – 211, Groundwork.)* Chamber – Yeah! I know.

Regarding the Barstow project, my designs were well received according to Gale Kenyon the city's Chief Engineer. Barstow had to find the money somewhere and that took so long they probably made other architectural arrangements. Probably they decided not to take their City Engineer's advice having an out-of-towner do the work. If they acted on any of the projects, they probably used a local architect. I never heard from them again. It's OK with me. I was not into jails anyway

You are right about the PBY. It was able to fly slow and under radar's detection and was used for reconnaissance.

Yes Viveka, my oldest daughter is studying a combination of Eastern and Western medicine. Her goal is to be a licensed acupuncturist. She has finished almost four years of intense schooling and will be doing residency for a year at the hospital-school called *Yo San University* in September. She is apprehensive about doing acupuncture for the first time on actual patients. I encouraged her by telling her when I first started my own business I had severe panic attacks, too. She will be 41 in September and I'm very proud of her for attempting so difficult a profession at so late an age. Marge's daughter, Marggy, is married to Mark Wheeler and they have one five-year-old son. This is the family that raises organic herbs in Southern Oregon near Grant's Pass. So the rumor goes, their farm is the second largest herb farm in the country. It is called *Pacific Botanicals*. Unfortunately, Marggy and Mark are divorcing. There is much hurt, but they are attempting to divorce amicably. I tried mine amicably, too, but it didn't work out. Of course Marge, mother of Marggy, is upset. Their farm is large, beautiful and well kept. Marge and I feel both of our families have lost a beautiful dream. For the present Marggy and Mark will both operate the farm.

That's it for now. I was delighted to hear of your trip through the Panama Canal. Late last year I finished David McCullough's, *The Path Between the Seas*, which is a book about the building of the Panama Canal. I recommend the book. It reads better than a novel. I would like to go through the Canal myself sometime. Marge and I are going to take our first cruise ship from San Diego around the end of Baja California and up the Gulf side to Loreto, then back with stops at La Paz and *Cabo San Luco*. You seem like a lucky dog to go on those fabulous cruises to Hawaii and the Canal that would cost an ordinary person an arm and a leg. Of course you have to work. But then, you enjoy your work. Dinner's ready!

Best wishes. Glad you liked the book. I enjoy your letters they're the next best activity to visiting.

Sincerely,
Doug

5-12-05

Chuck,

Your letter and photographs of your own work blew me away. I thought the photos and information on your work was terrific! I don't think architects share their work with each other very much. I appreciate you sharing what you did with your life and architecturally with me.

Your own house *(8 photos with explanations)* is clean, sharp, and contemporary. It looks like a great living environment. Did you lose it with the divorce? Does Mary Ann and Nancy still own it? What happened to it? I can see you like to take doors and windows all the way to the ceiling. Sometimes I think if designers and architects knew this one idea, buildings would look cleaner and better designed. *(When you put in a door, you design the wall. I don't think many people know that.)* I liked the sliding birch doors between the dining room and family room-kitchen. I have used a lot of clear red birch in my buildings, though we can't get it anymore. You and I must have used it all up. I liked the similar floors throughout the building. Were they all cork tiles with a vinyl finish? I also liked your use of indirect lighting behind a valance. I do a lot of that. I've learned a lot about lighting from doing lighting and working with lighting experts - if you can see the source of the light that's a no-no. *(Except for direct lighting for reading and close work.)*

The J. C. Wright house in Edwin Park is sensational. It has those beautiful Neutra-like lines. If I'd done it, I'd have it on my office wall. Thanks for including a plan and sketch of the atrium. *(Which can be seen from the outside in one of your exterior views.)* The circulation running through the house around the atrium works wonderful and your use of glass and clerestories makes for an open, free house. I'd say the stone wall is pretty powerful by itself, forming a strong anchor for the house.

The *Englewood First Church of Christ Science* is also a fine work. The colored photos gave me a good feeling of the interior warmth of the work. It has a residential feeling. In fact with the strong walls, open beams, exposed plank, compatible tan carpet, it almost looks like something I would do in my residential work. I would feel quite at home in such a structure.

Which brings me to a philosophy I have been

thinking about recently: a structure in conflict with itself. Who'd want to live or work in a structure battling itself? What is a structure embroiled with inner conflict? It's an *unharmonious* house. How do you design a house at peace with itself? Of course, make it *harmonious*. How? If a structure is a background for people who do activities, make the fewest number of simple elements you can invent: one kind of walls flowing inside and out, glass dividing the inside from out, one kind of roof flowing in and out, one kind of floor inside and out, etc. Then bring only things that are loved into the house. How could I *not* want to come home like that? It's just an idea.

I loved the design and rendering of the Denver sixth *Church of Christ Science*. It was a beautiful work. I'm sure it was far better than what the other more traditional architect did. I loved your clean, clear rendering. I wish I could draw as well.

About the *Western Farm Bureau Life Insurance Company* building: I have a saying, every time I'm disillusioned, *I gain maturity*. I guess you became *mature* on this building. Another: *No good deed ever goes unpunished.* The design is great. I hope the new architect built it that way.

Berthoud's bank and urban renewal shows a sensitive design. Other than your excellent designs, you learned how to draw parked cars and trucks.

The small office for the *Ajax Sweeper Company* near the *Bronco Stadium* looks strong with an attention to harmony and strong use of good materials.

The houses for Hazleton and Price look terrific. I like the economical approach while still doing something out of the ordinary. I designed a number of houses of this type in the past. I liked it! They didn't take as long, and they went up fast.

Enough about you, let's talk about me! I want to say you have completely understood what I was trying to accomplish in *Growing Edge, How to have a personal life and also be an architect*. I like one of the best compliments I've received as a personable gift: *"I rather think of it as the presence of Divine guidance lifting and protecting this remarkable individual who remains humble and grateful for each step along the way."*

To understand the title *Growing Edge* it is important to consider the two previous books. *Early Stories* told of life from birth through college. *Groundwork* tells of doing basics in architecture; getting experience, getting licensed, trying a partnership and earliest work in Malibu. *Growing Edge* is intended to mean the growing edge of a family and an architectural practice. The fourth book will probably be called *Malibu Practice* in which I will tell of a later, more experienced architect employing three or more draftsmen and having a multitude of jobs. The title *Growing Edge* essentially means my growing edge in architecture, but also applies to my personal life. It takes place in the time we had our third daughter and were on the growing edge of our lives in Malibu.

If I had located in Denver with the cold, wind, snow and different climate, I would, of course, had to take all those conditions into my thoughts to make indigenous houses. As it is I only have to take into context California conditions. Building in a city like Denver would certainly affect my designs. My philosophic viewpoint would change with the climate. In fact, when I built a house on Kauai, I outdid the Hawaiians with a beautiful what-you-see-is-what-you-get pole house.

I appreciate the "*I — me*" criticism. It is well taken and it will be corrected in the next printing. *(If there is any.)*

Regarding rubbing pencil graphite on the back of tracings and erasing for the lighter affects, I don't think I learned this technique from you. I don't remember doing any preliminary designs at Grove's office, nor was I capable of making any. As I remember I learned this technique from a draftsman for whom I did beautiful work in 1957 named Dick Irwin.

Regarding Linc Jones. *(Architect, Lincoln Jones.)* I lost touch with him many years ago. He was kind enough to send me a hundred dollars to help me recover from losing our house in the 1970 fire. I have no knowledge of Linc since about 1976 when I dropped by to see him on my way to visit my parents in Golden. I met his new wife and saw his new house in Steamboat Springs. We horsed around as usual and I had a pleasant time. I'd love to know what happened to him. He seems to have just disappeared.

Regarding the demise of the pedestal house: I'm already fifty-pages into the fourth book. I'm starting it off with the article I wrote for the Los

Angeles Time's Home section that describes how it feels to lose a dream house. In other words, I'm burning the house down in the first part of the next book. I figure you have to start low and downtrodden so I can gradually regain an office and build a new house over the old foundations. *(Alas, I recover)* That should play well for book 4.

Regarding the need to relocate: After 4 years of living in the pedestal house, Karon, the kids and I were tired of the steep hill. The house originally was designed as a spec house for another family. We wanted our *own* house, designed for *us* on a flatter piece of ground and one nearer the beach. In the long run, economics would not allow this. Our minimum budget forced us to rebuild on the same site. I'll explain in the next book.

Regarding Marge's approach to writing about my former wife: the writing about each of our former lives is a non-issue. Neither of us can deny we had former lives. Jerry, Marge's first husband of 25 years and the father of her four children, lives close by and I see him at family gatherings. We are friendly and at relative peace with each other. Karon, my former wife died of cancer in 1987. Marge is a Marriage Family Child psychologist as well as an Art Therapist and has learned to accept and roll-with-the-punch on most human behavior. She is doing her own autobiography in 20 or 30 large PrismaColor drawings of her former life. She and Jerry are in them along with her young children. For the past 29 years I have been in love with a good woman and there's not much of anything that bothers us.

Chuck, I appreciated your letter more that words can tell. I loved the fact that you shared your former work with me. It's great and I can see we are on the same page. Hope all is well with you and that you are doing the things you love to do.

Love and best wishes,
Doug

12-21-05

Dear Chuck,

You wrote the word, *(Procrastinator?)* after the date on your last letter, which was on June 17, 2005. Will the *real* procrastinator please stand up? I'M IT! Sorry I'm so late in answering, but I have been late in answering all my mail lately. I've been busy. Kristofferson is finally in the Malibu Building Department. The Building Department says with Coastal Commission approval, a property with no sewers that might have been slide-prone, *(but wasn't)* has got to be one of the most difficult of Departments to get a permit. Though I will be 78 in 10 more days, I'm busy. Next year I will be doing work in Santa Barbara, Rancho Palos Verdes, Long Beach, Santa Monica, 2 remodels in Malibu, one new house, *(Kristofferson)* and responding to 4 or 5 calls from the recent, but rare publicity Marge, and letters I have received from my publications in the Los Angeles Times. *(Enclosed)*

You mentioned in your letter that subsequent inhabitants of your buildings messed them up. I had one buyer of one of my best beach houses *(constructed of clear heart redwood when redwood was not on the endangered list)* called me to brag about how he had improved my house by painting the entire exterior battleship gray. I said *"Oh! You did, huh?"* He never heard of the rule *"truth in materials."* Needless to say I have not had much contact with him since.

I'd say fifty percent of my constructions are irreparable, 35 percent may be salvageable, and 15% are still in good shape. You have to *maintain* a house. I had one client who owned a lumber company and had his house built of Cedar. He decided to maintain it for the first time after 35-years. We almost had to do it over. Let's make up a song and call it *Architect's Lament*.

It sounds interesting that you're dancing on cruise ships. Marge and I went on one cruise ship and enjoyed dancing and entertainment. I will send you something I wrote after it was over. If it's a repeat of something I've already sent, throw it away. Golf is interesting and dealing in the stock market sounds scary. If you can make money, more power to you! About mudslides: I feel for the families, but am usually overjoyed scenery has been improved. Some of the houses that slid were pretty bad. Except for the tragedy of the families from an artistic viewpoint it was good to see them go down the hill.

For the past two or three years Marge has been concentrating on her artwork. She works in PrismaColor pencil in her studio for 3 or 4 hours daily and is quickly coming up with enough for a one-person show. I know she will do well. Her work is meticulous, has a point and is meaningful.

Recently our house was photographed for the L.A. Times. Marge and I were interviewed by next year's assistant editor of *House Beautiful*, Barbara King. I also realized it was a good public relations opportunity, so I synthesized some of my views into *"sound bites"* and gave them to her. I think she did a good job with the article. The ideas I expressed are not mine, but are philosophies I've heard over the years that expressed what I believe. I remembered them and collected them. It's surprising how in Southern California, even the term indoor-outdoor living is not commonly understood.

Sorry to cut this short, but I'm staring at 45 more Christmas cards to write and it's the 21rst of December. Please write and we'll talk again. Or send a Gmail.

Merry Christmas,
Doug

4-20-07

Dear Chuck,

Got your letter of April 13, 2007. Your call came in on my answering machine on what you thought was my birthday, March 24th 2007, however that date is my brother Dave Rucker's birthday. My last one was December 31, 2006 when I was 79. I hasten to say I was not put off in the least, but honored and greatly surprised you'd attempt to remember my birthday at all. Come to think of it, I don't know *your* birthday. What is it? Maybe I'll remember it and give you a call, though I'm not sure I trust myself to do that. I'm happy to be alive at all, these days. I've had a pacemaker for five years and should have been dead long ago. I'm pleasantly surprised every time I wake up. It's beautiful here. Marge and I often think we've died and gone to heaven.

The small house *(1,925 sq. ft.)* for Kris Kristofferson and Lisa has just been roofed with fiberglass shingles and is beginning to shape up. Stuccoing will begin on Monday. A picture is worth a thousand words, so look for a picture with this letter. By the way, we did run way over the budget, but then this *is* 2007. A client can write a program *guaranteed* to run over the budget. That was done in this case. Lisa wanted to spend $550,000.00 and it ran $950,000.00 and we were able to cut back and it is now $750,000.00. Where it will end up is anybody's guess. So far I have to believe they still like the house and me. They just recommended me to *"Spider"* Giraldo and his wife Pat Benatar. I was glad Lisa and Kris wanted to recommend me, since they are always on tour, it is sometimes difficult to know where I stand with them.

Last week they were in Nashville where Kris picked up the Johnny Cash Visionary Award joining previous winners, Hank Williams Jr., Loretta Lynn, Reba McEntire, the Dixie Chicks and his late friend, Johnny Cash. I now know what an absentee owner is all about? The Kristofferson's are in Nashville, Shreveport, Chicago, New York, Europe, home for a few days, then off to Australia, then Canada and so on. As far as I know neither Kris nor Lisa have seen the house since construction began. I'm handling the job through their accountant who sometimes acts as a go-between. I make occasional cell-phone calls to Lisa and send faxes and letters. It still may be a good house.

On other jobs, I just got a permit and will start

supervision on a $350,000.00 addition in Long Beach for the daughter of a former Malibu client of mine. I'll have to drive the 405 Freeway, which is like racing at the Indianapolis 500 at eighty miles an hour. When slowing, I hope I don't get pounded from behind.

I know when the addition is finished it will be a good one. Also I'm refurbishing and upgrading a house I did 40 years ago. We've put in a long skylight, *(24 ft. x 2 ft. 6 inches)* new kitchen with blue-eye granite countertops, new appliances, new trellis, new bronze anodized sliding glass doors and windows, new glass handrail, etc.

I only work with two contractors. If a new client doesn't like the contractors, he can go elsewhere. I haven't lost any clients because of that. Also, two weeks ago I got a long distance call from an old friend who lives in England. He used to run the Malibu Blueprint Company at the same time I had an architectural office in the converted *Rindge Railroad Terminal Building* located across Pacific Coast Highway from the ocean. I hadn't seen him for 17 years and though at earlier times we were fast friends, three weeks ago I couldn't recall his name. It turns out he still owns a house high on a hill in a newly expensive neighborhood overlooking the ocean and Point Dume. He wants to do a master bedroom, bath and decks over the garage and trusts me enough to send me a check for $20,000.00 to open an account in my name so I can write his bills. I'm doing as-built measurements now and have ordered a 2,500-dollar partial boundary survey with contours. Will wonder's never cease?

I still enjoy snail-mail. Somehow, I think it's more personal. This summer after the snows you can get back to your golf game. What are you shooting these days for 9 and 18? The puzzles sound exhilarating - if not exhausting. Send me a picture of the completed artwork.

Regarding my latest book. I haven't written anything new for about a year. I do have 60 pages done on volume 4, in which I begin by burning my house down. I felt if I start low, perhaps I could build to a positive climax. *(My divorce? No! Meeting Marge! Yes!)* Even though I can't wait to get into that book because I'm working too hard, I was nevertheless able to put together 60 humorous essays written over the past thirty years in *Book of Words*. I'm looking into the Internet to have it printed. It is said books are good looking and done inexpensively.

When I have them printed, I'll send you a copy. *(No pictures.)*

With just a letter to reestablish our friendship and communications, I write Charlie Davis, an old architectural buddy I met during my 4-1/2 years at the *University of Illinois* who is 86. I also write Don McGregor, an old hitchhiking buddy of the late 40's who is 80, has diabetes and has lost both legs below the knee. My old buddy, Roy Scheck, took a swing at a golf ball, made a twisting follow-through and fell on the green grass with a fatal heart attack. That's probably a good way to go, though I'm not suggesting it. Chuck, stay healthy! Thanks for your Gmail address. I'm filing it. I'm sending my best to you and keep writing now and then.

Sincerely,
Doug

7-17-10

Chuck,

    I just thought you'd like to know the real truth about *How Colorado Got Its Name*. (I included a 5-page funny story naming Colorado.) Regarding Emails: I have no Email, but do have *Gmail*. I've had it for well over a year and have been so sorry to miss whatever you had to say. My Gmail address is: ruckerdoug@Gmail.com

    Regarding Obama. I'm still an Obama fan, having no better choices than Rush Limbaugh, Sarah Palin, or Republican talk show hosts, and though Obama may not have the polls he once had, I have to remember the impossible mess he inherited with six years of the last administration and the near-impossibility of overcoming that disaster in a little over a year. Obama will need more time. I'm also critical of religions and lots of things stick in my craw such as the oil spill in the Gulf, global warming, melting glaciers, advancing nuclear threat to all humanity, exponential increasing of the world population, overfishing, air pollution and the U.S.'s obese 7% of the world population *(US)* using 25% of its resources.

My last book, still in the proof stage, is called *Harold and the Acid Sea of Reality*, wherein, in more personal terms, I discuss all this. I'm waiting until I get more pressing things off my list before I finish. When it's printed and if I get to it at all, I'll shoot you a copy.

That said; now to the more positive things, *(if any)* in our lives. Speaking of Mortal Disasters: *(Just joking!)* I'm glad you're recuperating from a collapsed lung, gash on your head, knee replacement and cataract surgery. The cataract surgery, thank God, went well. Gaining weight is a good thing and it's fortunate that physical therapy is beginning to help your rotator cup. I bet your consistent exercise is helping the strength of your over-all body, breathing, digestion, elimination, etc. Do you feel better? The good things are you don't have to work; it's good to play golf and have the loving companionship of Smitty! Marge and I probably watch the same programs you do. *(Law and Order, The Mentalist, Castle, Burn Notice, The Good Wife, Royal Paynes, White Collar, etc.)*

Marge has been a dancer all her life and she and I met in a *Dance Improvisation Class* and stayed in it for over 10 years, so we're absorbed in programs like *Dancing with the Stars* and *So You Think You Can Dance*. Oh! Yes! We enjoy Tavis Smiley and his interviews with politicians, athletes, writers, movie stars and vital people who contribute positively to the problems of the world.

But when Sunday rolls around we take off from our weekly work: Marge from her one-day-a-week job at *Hospice of the Conejo* as Grief Counselor, her PrismaColor pencil artwork which she is very serious about and trying to make an art name for herself, her shopping, cooking, clothes washing and telephone calls, and I from my sometimes architectural work, essay writing and new artistic photographic work. If there's time, we go to the beach and points east.

We will investigate the sand at the *Channel Islands Harbor*, *Ventura Harbor* and *Ventura Promenade*. Easterly, we visit *Lake Casitas,* a large man-made reservoir and lie in the dappled shade of our favorite tree on a hill overlooking the lake, and lie on a blanket with pillows and take naps and read. Sometimes we visit Ojai and eat salads with other vacationers under the multi-colored umbrellas at *Antonio's Mexican Restaurant*. I have my favorite Chile Rellenos with rice and beans and a bottle of Corona in a cold glass with a slice of lemon.

On rare days we drive along windy Highway One and the breezy coast next to the beautiful blue Pacific and visit Santa Barbara where we eat at our favorite health food restaurant, walk State Street, which is full of young people in love, children and the elderly of all denominations. Sunday is taken with people of all sorts shopping, talking, walking and eating outside. Along the *Santa Barbara Beach* we visit the mile long outdoor art exhibition with booths and examine the variety of local artistic achievements. Afterward we walk to the end of the pier and watch fishermen, sea gulls, multitudes of happy people and pelicans flying V-shaped across the end of the pier or sitting, asking for hand-outs on the railings and roofs.

Other than that, during my spare time, which I'm happy to say, I seem to have more of, I'm getting into photographic art. At the moment I have three pieces in the *San Buena Ventura Artist's Gallery* at the *Ventura Promenade* on the beach. The theme is *In Your Face,* a show of self-portraits. And I have two pieces showing at the four-story Atrium Gallery in the *Ventura*

County Government Building in Ventura. The juried show with 3 prizes is called *"Whimsy"* where all 77 pieces of artwork are to strike a whimsical nature. I put in a photo called *"Mr. Universe"*, a picture of smirking me emerging from the black night with stars across my face under one of Marge's PrismaColor drawings showing the *Tree of Life*. The other picture was constructed of sixteen art-cartoon panels where the man and woman have an argument using squiggles for words, but it turns out all right. I'll send you the two along with an extra artwork, a picture of a dancing woman taken by telephoto lens at a *High School Dance Competition* held last year in Las Vegas. My copy is of a greenish tone and I'm going to frame if for the office. Hi to Smitty! Keep writing. I value my old friends.

Doug

8-10-10

Chuck,

Got your letter of July 24th. I've been having trouble staying on the Internet so if you've sent a Gmail *(not Qmail)* I've not been able to see it, or them. First I'm *ON,* then it goes *OFF,* then I have help getting on again, then it goes off. I've not been able to communicate with my daughter in Windsor, California or other friends, or even get information or buy an Internet book. I live in an area that doesn't have cable, so I'm stuck with having a wireless Internet connection. Because Marge has a laptop a hundred feet or so away, we have to have a wireless Router to handle both computers or pay $61.00 twice a month for individual connections. I'm not up for that. What I want is a router to handle both our computers that I can leave on indefinitely and not have to worry about it going out with all the difficulties and special help I need to reconnect. Chris, Marge's 24-year-old son and computer expert at *Pierce College* is coming over at 3:00 PM today to get me out of my Internet mess. So bear with me and wish me luck.

Marge and I have a deal working. She cooks dinner and I read to her. We only do a few pages at a time, but over the years we've finished some pretty long and interesting books. The one I'm finishing now is *Seven Plays by Henrik Ibsen*. Do you know him? He wrote *Hedda Gabler, A Doll's House* and *Peer Gynt*. Do you recall any of those plays?

Anyway, last night, while she was making Red Snapper, corn on the cob and co-slaw with pineapple and I was having my nightly *Eye of the Hawk* beer, I read her your letter. She liked it and when I finished and said to her, "*Chuck can really write, don't you think?*" She replied, "*Yes! It sounded fine.*" So, I got to thinking I'd encourage you to put down your thoughts in a readable way. I'm a reader and I'd read them. It's possible you've been doing that right along. I don't know. It doesn't take much and you have a feeling for expression. Just a thought, but write your essay on your operation experience. It's good practice.

Regarding retirement: For some reason the word retirement scares the *ZZZ* out of me. It seems the next step after retirement is death and I still have lots for which to live. As long as my pacemaker keeps my heart beating steadily, I should be all right, right? I live the grateful life because I've got a paid for great property and a wonderful, intelligent and artistic wife, and enough in savings I think will last - depending on how long I live. We also have 7 children all out on their own with eight grandchildren, the oldest of which is 19 and the youngest, 7. They make life worth living and I feel they need us for a little longer. Life is good. I'm hoping it doesn't change.

Moneywise: In June of 1993 I had a minor stroke and for a while I was a little out of it. I didn't have much business, so I relaxed on the couch and read a lot. To get exercise, I decided to hand dig foundations for a separate architect's office on the side of my hill. I worked for months, recuperating physically from my stroke and picking and shoveling and wheel-barrowing until I got to the place where a contractor friend could pour footings and a slab. I had no idea how I was going to finish the structure, but I did have it designed and did have a permit.

Then late that year Malibu burned down! Over a hundred houses were lost between Topanga and Malibu and I lost 6 houses in that one fire. After a short time it became evident that Malibu was going to need new houses and three of my former clients, owners of the houses I'd designed for them, called me with intentions to rebuild. I called Ray Fontaine, the contractor I'd used on several remodels before and the one who'd done my own 710-square-foot-house and said, *"Ray! Build my architectural studio!"* He did, and within three months it was waterproof and ready to be used. I moved all my equipment down from our small house, wherein had been not only my working architectural studio but our only home. By that time a few of my old and new clients were ready to begin work. In 1994 I immediately had five expensive houses to do with insurance money: the Munro house 3,200 sq. ft., Craft house 2,800 sq. ft., Bright house 2,200 sq. ft., Jones house 2,000 sq. ft. and McCarty house, 6,000 sq. ft. I did a very thorough job on all these houses at 12% of the cost of the work, and though I had my head down and my elbows up most of the time over the next three or four years, I was able to put together a savings account earning the lesser but safer insurance company interest. Marge brings in a thousand a month because she's still working. This helps a lot. I have consulting and drafting work upon occasion and can contribute to our income.

Then, of course, Marge and I have Social

Security, without which we'd be in trouble. We think with a paid-for house it will be a reasonable amount to live. With substantial savings and Social Security, we'll be able to make it. That all depends on how long we live and how much living costs go up. You and I are probably both in a similar boat.

How was the *King Tut* experience? Did you take the guided tour and see the 3-D movie? My two high school buddies are both dead. Roy Scheck became rich, lived on top of his own 13 story Florida penthouse, raised six kids and was on the course swinging a golf club with a great follow-through when he fell to the ground dead with a massive heart attack. Don McGregor, an outdoor high school friend who was married with two kids, died in the hospital due to a heart attack having lost two legs below the knee due to diabetes. Yes! When you get old, it's no fun to lose your friends and any health you had. We're both lucky.

I now have five photographs on display in two galleries. One in the *San Buena Ventura Artist's Union Gallery* off the *Ventura Beach Promenade* and the other in the 4-story *Ventura County Governmental "Atrium" Gallery* in downtown

Ventura. Last Saturday, Marge and I and Viveka and her husband, Tom, went to the opening. The artwork was contemporary and sensational. My work held up superbly and we made some new friends. One artist was also a writer and had just self-published his book. I bought one for 20 bucks and it reads like wild fire.

Hey! I'm running out of gas, so I'm going to say, *"Till next time!"*

Best wishes,
Doug

1-5-12

Chuck,

Got your New Years letter. My art celebration was well received. I sort of got my first show on a fluke. Marge, a fine PrismaColor pencil artist, was invited to have a show of 25, or so, pieces. She didn't have enough stuff and the person handling the show seemed desperate, so I said what about me? The handler said great and scheduled me in. We had about 30 people, some that I didn't expect to show, and some I expected that didn't show. The show will be up until the 18th of January.

Through studying large and small galleries and subscribing to photographic and art magazines that handle art-oriented photography, I've discovered, when it comes to art, there are very few working on a continuing basis into *Digital Reflection* art photography. It looks like I've found a *"shtick"!* I have lots of highway junk I've found as a subject and a reasonable artistic eye discovering photographic subjects reproduced at a larger scale as art. Besides the 27 pieces in one show, I have two pieces in two more galleries and two more going into another one this Saturday. Also, I've one that was selected and is on display at the *Montecito Bank in Ventura.*

Sales? I've sold two at $300.00 and $125.00 and have another my son-in-law wants. I've agreed to let him have it for the price of the frame - $125.00. Yes! There's a lot of money in reflection photography. Just kidding.

Marge has two wonderful autobiographical, PrismaColor pieces *(16" x 20")* in one gallery and two larger cloud pieces in another *(30" x 40" and 20" x 24")*. There is a world of difference between each of our works and you'd have to see them to know.

Thanks for your statements about my continuing architectural practice. It doesn't occupy much of my time at the moment. So far, I've been extremely fortunate in having clients become friends. They say their lives have been changed for the better by living in one of my houses. If the truth were known, according to the building department, houses on the average turn over every seven years. I have done about 90 new houses and about 50 remodels. There have been lots of my houses that have changed hands and when that happens, the new owners are less sensitive and rarely maintain the house or remodel it as originally designed. Sometimes they just paint the sucker white or blue or a multiplicity of colors, which, of course, drives me mad. Such is the fate of an architect who cares about his work.

I've come up with a new phrase at 84 years old. *Let the scroll of life continue to unroll.* That seems to be happening for both of us and there doesn't seem to be much we can do about it. We are who we are!

I am sorry about you losing your daughter. I can only imagine how difficult that must have been for you. You have my sympathy. I'm pleased she had lots of friends and that her life was celebrated. When someone dies, they've lived their life completely - from beginning to end. Nobody gets out alive and at my age, the scroll is more than significantly unrolled.

Oh, yeah! Two months ago, on my 5-day-a-week walk of almost 45 minutes, I climbed a broken chain link fence at the end of nearby park to see the fantastic view of *Agoura, Westlake Village, Thousand Oaks,* and parts of *Camarillo.* Our house is at an elevation of 1700 feet above sea level and the mountain at the end of the park drops off about a thousand feet into the rich valley of foliage, homes, freeway and towns. On the way home, back over the fence, my heel clipped the top of the fence, I lost my balance and fell four and a half feet almost flat on my back. My elbow dropped to brace the fall under my back, and I'm sure by the size of the hematoma, it helped break my rib that punctured my left lung.

It was still a long way back to Mulholland drive: a 15 minute walk along dirt paths, past the pond and parking lot to where I was able to flag a car driven by a wonderful, sympathetic female person. She took me on a 20-minute ride to the Malibu emergency hospital. My regular Doctor took x-rays and pronounced I was going to live. The first month was difficult, but the second is better and the Doc thinks I'll be completely healed in another month. I hope so. I heal minimally every day. Chris, my stepson lawyer, says, *"You're going in the right direction."*

Chuck, we have always valued each other as close friends and influenced each other's lives. You've gone down as an important personage in my autobiography and it is my wish for it to remain that way. Politically, we are on different sides of the fence and I'd like, in a respectful and friendly way, for you and I to agree to disagree. Your political views *do* upset me. I usually delete them because I've heard them before and almost always feel they don't include the entire picture. If you'll agree to send me no more political viewpoints, we can continue our mutually valued relationship. Agree?

That said, I valued your reasonable letter and do value our friendship and sympathize with the loss of your wonderful child. I think you are a terrific person and though we may not agree on everything, I'd like to continue our friendship as it has been in the past.

Appreciatively,
Doug

3-14-12

Chuck,

I can't believe it's almost three weeks since I got your letter of February 25th. It was a complete and very informative 2-pager letting me know about many of the things you've been doing with your life. I think I remember you being in the *Navy Air Force* and afterward becoming a flyer and flying all over Illinois - no mean task.

In the summer of 1949, just before my last semester at Illinois, I worked with two surveyors and a college Independent House-mate surveying five man-made lakes all over Illinois, Springfield, Kankakee, Peoria, and I forget the other two towns. We mapped new and shrunken shoreline and silt-laden depths for the *Illinois Department of Water and Power.* Lakes over the previous years had silted up behind the dams and the Water Department had to know how much in order to take remedial steps. I worked in Levi's, gym shoes and no shirt and by September at the beach I was a young man of two colors, deep tan - almost black - on my back, chest and shoulders and a pasty white on my legs from my belt-line down. With just trunks, it was embarrassing at the beach.

You had terrific architectural experience doing houses, schools, offices, apartment buildings, and manufacturing plants. I stuck mainly to houses and additions. I guess my philosophy was *"think small"*, however I did do four apartment buildings, an eight unit in Santa Monica, a twelve unit in Malibu, a four unit along with a 6,000 sq. ft. *Recreation Complex* and pool for Malibu's largest *Mobile Home Park* overlooking Zuma Beach and the ocean, and another eight unit in Isla Vista adjoining the University of Santa Barbara.

It is a good thing doing jigsaw and crossword puzzles. I've heard they keep your mind agile.

Regarding golf: I'm still in touch with my high school friend, Jim Baer, who also likes golf. He plays whenever he gets the chance. Jim and I ran on the track team and Jim played left halfback and I right halfback on the *Chicago Austin high school* football team. Those were in the days of the T-formation first developed by one of the four horsemen, Frank Leahy then coach of the Notre Dame football team. The Chicago Bears soon followed suit with the *"T"* formation along with Bill Heiland, the *Austin High School* coach. Anyway, golf's a good sport, though I never seemed to get into it. My only experience with golf was with you when we used to hit the course for nine before work at Groves. Those were the days! If your game's not improving, let's face it, getting old sucks.

I knew you had two marriages, but I didn't know you had three. My first wife died of cancer after we were divorced for five years. It was a long illness and it took its toll on the kids who were more closely involved. It was a sad and horrendous time for all of us.

I saw your Gmail on the Concordia. It's hard to believe the amount of detailing that went into that cruise ship. I guess that special attention goes into most cruise ships, but you'd know better than I, having enjoyed and taught dancing on so many. A real tragedy, the Condordia, not only for those lost, but for all that hard work and human constructive and engineering endeavor.

Hope your TV screen is working properly. We live in the mountains and have no cable and so have a satellite dish. It's been working well for twenty or so years without being touched even though we have the strong Santa Ana winds blowing mostly in the fall. Its basic costs are about $60.00 per month. We are not buying more "extras." Marge and I also spend about the same length of time as you two watching TV. We like the NFL/AFL games, *The Mentalist, Castle, Blue Bloods*, Gary Denise and Melina Kanakarades in *CSI*, and *So You Think You Can Dance* and *Dancing with the Stars*. But besides TV shows, I frequently go to the County Library for movie DVD's. If nothing's on, we put on a library movie. Do they have free movies in the Library in Denver?

Your trip to see the musicals seems like fun. Marge and I rarely like to drive into downtown L. A. to see those, however, we did go once many years ago to see the *Lion King*. I liked the costumes and special effects. It reminds me, when I was first married, my then wife, Karon, got me into a Gilbert and Sullivan production at the *Pasadena Playhouse* for a month called *The Gondoliers*. I sang in the chorus and did dancing which was a real stretch from being a former football player. I was also in George and Ira Gershwin's *Girl Crazy*. I sang in the quartet, led the dancers and sang in the chorus. Also, we sang for a month in the *Downey Playhouse* in the opera by Giuseppe Verdi called *Rigoletto* - a great introduction to the classics.

So, your life has been fine and good with lots of human experiences. I've heard it said, *"No matter when anyone dies, by the time they do, they've lived their whole life."* I guess we're lucky to have lived 84 - 85 years. Many of my old friends are already dead. That puts us on the line. Are we going to worry? I'm not worrying, but am trying to get the most out of every day I have. I heard a good saying that's given me confidence. *"Let the scroll of your life continue to unroll."* For some reason, it makes me feel better and I'm going to do just that.

I'm pretty much over my broken rib and just have pain when I walk in my left knee and with long walks, pain in my hips. Yes! I'd like to feel young again, but then, what choice do I have? The best is *"Live a grateful life."* Guess I'll do it. Marge says hi!

'Til next time, be well,
Doug

1-15-13

Dear Chuck,

   Glad to hear you are still moving under your own power. I enjoyed your letter, particularly about what's going on in your life, such as sewage coming up through your shower saturating your carpets, and removal and replacement of your toilet and your car door *"banging"* fiasco, gas cap and car keys adventure. I'm delighted you are finished with all that and glad your precious elderly years are filled with such meaningful and orderly business. I'm also pleased that now and then you're able to get out for golf. I know it may be a moot point, but what's your score for 18 holes these days? The *Shark Rug Shampoo* machine handle must have been a thrill.

   I'm glad your health is good, but blanched when I read each lung collapsed at different times. That doesn't sound good. What does the doctor say? Regarding implants; Several years ago after a lifetime of work, a couple of cavities in my teeth next to each other renewed themselves and my dentist suggested I have them both pulled and at the same time have two new teeth implanted. They were to go into the upper left jawbone, so gathering courage, I made an appointment with another dentist who does that sort of thing and sat there with my mouth open for an hour while the dentist and a couple of his assistants drilled for oil.

Eventually the ordeal came to an end with titanium anchors for two teeth firmly affixed. At the same time they installed two temporary teeth. I was to wait six months while the jawbone more solidly adhered itself to the titanium and then my own doctor mounted matching caps. My implants are my best teeth. I can gnaw on bolts of iron and chew wood. In other words, I haven't been sorry. In fact, I've been delighted. Though I'm not a rich man, the $3,000.00 price has long since been forgotten while I'm still enjoying new teeth.

Regarding investments: A few years back Malibu again burned down and I became busy for a couple of years putting everything back. From this blind luck, I was able to put a few bucks away, so with our joint Social Security and supplemental amounts taken from interest on our savings, we're able to live a frugal and satisfactory life. Of course, neither of us knows how long we'll live and that's really the crux of the matter. If we both die tomorrow, financially we'll be OK and the kids will get something. If we each live another five years, what with inflation and all, who knows what our financial position will be. I always think an option would be living off a *Home Equity Loan.*

Can you do that? Can we live in a 710 sq. ft. house with separate studios? Mine - about 500 sq. ft. Marge's - about 384 sq. ft. Thankfully, our property is free and clear. Though our mail comes through the *Malibu Post Office*, we're about three miles inland from the Malibu City line and we're actually in *Un-incorporated Los Angeles County.* The property may have changed a little since you and Smitty were here. We've added a new carport and a 12 ft. by 24 ft. concrete block storage area with growing landscaping.

About investments: Call Dave Rucker. He's in Golden at 303-279-2783. I know nothing about investments. We have a man whom we think is trustworthy and he's put us in safe, but slow-growing insurance company funds.

Regarding puzzles: I have to hand it to you taking on a 1500 piece jigsaw puzzle! I seem to have lots of thoughts to get down before I die and am writing my opinions for everyone to see.

I've just finished writing a book called *Harold and the Acid Sea of Reality*. To quote from the back cover of the book, *"There are thoughts, philosophic, humorous, psychological, nostalgic, digressions and everything from God's final word to the ridiculous."* Few people read them, but from time to time I read them myself and it's fun to know what I formerly thought. The next time I print, I'll send you one.

Regarding friends and relatives who've passed away: I just lost a cousin this year. His name was Gene Kennard and he was 86, married to Johanna Kennard who had Alzheimer's for the past 4 years. They had five big and beautiful sons, but while coming to the end, I think Gene was done with life. I lost my first wife to cancer in 1987. Her ordeal was particularly brutal. I lost a dear friend at the age of 83 who went blind before he reached the end and John Marquis, my University of Illinois roommate, is on dialysis that can't be fun. He's been particularly quiet for the past few months.

My best architect friend, Rick Davidson, who was like a California brother to me, died in 1999 of cancer at 69. So I consider these last years of my life precious. During this time I find myself impelled to move forward with what I want to do with my life. It seems to be art and writing. I've shown 200 or so fine art photographs in about 10 galleries over the past three years and won 8 awards. Not all first prizes, however. It was more like *Honorable Mentions*. I'm continuing to show in the local area and just two weeks ago have discovered a really ugly wall, a culvert, really, on the ocean, under Highway 1 up the coast from Malibu. It has a dirt floor passing under Highway 1, a six-foot ceiling and dirty walls with graffiti written all over them. I took pictures of them with my flash camera and brought them home to work with them on my *Photo Shop Program*. I quickly made a book and told Marge, *"I whacked off a tunnel book!"* I made a proof copy with color pictures and black text. It's now at Kinko's to be spiral bound. After I proof it, I'll send it off to *Lulu Printing Company* and they'll make as many copies as I want for about $35.00 each. I was going to throw away the enclosed copies, but I'm sending them to you anyway. There are 24 color prints in the book that's called OFF THE WALL. Some of these copies are also in the book, but it will give you an idea of what the book might be like and what can be done with raw graffiti.

Regarding the fine, complimentary things you said at the end of the letter. It's a good thing if I made a positive difference in your life. Thanks for telling me. As well, you made a positive difference in my life. I was a social nerd at the time and though I've done pretty well for a nerdy guy, I still seem to have retained my nerdy-ness. Marge never achieved sophistication herself and that was what was lacking in her life. The nerds met each other. Though with four kids, she was a married partner not so sophisticated.

So, you can see, it all worked out for the best. You always met me directly as I was and never looked down on me for my naivety or un-sophistication or lack of social knowledge. I needed the trip to the Shriner's convention and needed to begin learning the ways of the social world. You were there for me and I haven't forgotten. Of course I married the first woman with whom I made love, had three daughters and when life told me *(after 25 years of married life)* things were not working, I found Marge and that made the whole difference. Our new seven kids are like one family. Thank you, Chuck, for remaining my good friend.

Sincerely,
Doug

10-12-13

Dear Chuck,

    Good to get your letter. Regarding reflections; I had written something for myself a few months back and I thought I'd send it to you. It's called *How on Earth Did I Get Into Reflection Photography?* I've also got something on Photo Shop, the text of a short color book I've called *Off the Wall* was completed on the spur of the moment. When I get it printed, which should be before Christmas, I'll send you a copy. In the meantime, here is the text *without* pictures. You have a few of the pictures I sent you, so you'll know approximately what the book will look like. *Off the Wall* discusses *Photo Shop* and it's uses and possibilities.

    You're right! *Picasso, Dali, Utrillo, Rouault* and *Van Gogh* have their own abstract qualities, but they're OK the way they are and with them, I'd not like to fuss. I'm doing something now on *"shards"* or broken pieces of headlight from a former fender-bender that happened on Kanan Road. I go for a 40-minute walk five days a week and find this kind of stuff. I found a 20-dollar bill last month and a dollar last week, but don't find money often enough to make a living.

Regarding your health; I'm pleased it is generally good and that you can breathe since they patched up your lungs. By the way, how do you patch up a lung? I'm sure it's not like patching a tire. I had a four-foot fall on my back trying to scale a fence two years ago and broke a rib that punctured a lung. It was painful for many months but eventually the bone healed and the lung repaired itself. These days, when I get tired the place where the rib was broken begins to hurt. Resting helps and soon it goes away. I should tell you I've been taking 200-mg of CoQ10 for the past ten years since I got my pacemaker. It's supposed to protect the heart, is an antioxidant and aids the immune system. Ask your doctor about it.

Regarding the dentist: I'd recommend dental implants. Do it once! Do it right! Have it done and after six months, breathe free. Believe me, you'll forget the cost. It's a lifelong thing.

It was good to get your intake on the reverse mortgage. We're glad it's working out for you. Marge and I are seriously considering the same. Our savings are adequate as long as we don't live too long. That's like praying for a short life so your savings don't run out. Our theory is unless you've got a better one, to continue to use our meager savings until they run out, then go into a panic and scurry around for a reverse mortgage. I'm not forgetting we have 7 kids and 8 grandchildren. Some are rich and others are poor. We're in the medium category. When the chips are down perhaps our kids can chip in. *(Don't count on it?)* Yeah! That ran through my mind as well.

Yes! I'm continuing my reading. I love to read and do so daily. I usually get my books from the library for a buck. Right now, I'm reading a book called *Marcel Duchamp*. Contained therein, *are stories about John Cage, Robert Rauschenberg, Jean Tinguely* and *Merce Cunningham*. It's a well-written book that came out in 1965 but still has stories of which I've never been aware. I just finished a book called *Storycatcher* that explains how all education started and continues by people telling each other stories. Naturally I like that idea and do so myself. Consider *Harold and the Acid Sea of Reality*. I have read two of Obama's four books, *Dreams of My Father* and *The Audacity of Hope*. The first, *Dreams of my Father*, was masterfully written.

Regarding my portrait: As far as acquiring

it, I'll let that one slide. I could have had it for $500.00, but I guess all I really wanted was the publicity, and that was only marginally acquired. By nerd I meant naïve, unworldly, inexperienced in the way things are done. I didn't drive until I was 22. I didn't know how to date until I met my first wife at 24. For a long time I felt inadequate as a man and threatened by apparent demands of the world. During my 25-year first marriage up until I met Marge, I thought of myself as Doug Rucker, "*boy*" *architect*. I feel lucky I made it to this age and do feel at this point I've achieved some kind of maturity. At least Marge and my kids and her kids treat me that way. In any case, I always felt respected and valued by you. Hence, see how well is our continuing relationship. I saw you as a much needed big brother having *"been there and done that"* showing me the ropes. I still feel and value you that way. So here is *Harold and the Acid Sea of Reality*. Let me know what you think. I've got *Book of Words, Where's the Cookies At?* And so we have *Personal Journey* waiting in the wings.

                                      Appreciatively,
                                                Doug

11-6-13

Dear Chuck,

Got your letter of the October 30, 2013 and was really taken with the thought of you using your *"last"* Listerine mouthwash. Was it an omen or premonition? Then you moved right into waking up the next morning *"dead"*. I love it! What an opener! That first sentence is the beginning of a highly readable story. I'm inflamed with curiosity! I just finished a book called *Storycatchers* by Christina Baldwin, a book showing since the beginning of humanity how it's the *stories* we tell each other that have educated people throughout time and have moved us out of the dark ages. Waking up *dead* is a great opening line and the following story is one that will definitely put humanity on the move.

The heck with what *"ought to be done"* like *premise, satire, intrigue, predictable result, surprise.* It's not a satirical story. Your story has what all stories should have - *unpredictability*. It has both *surprise and intrigue*. Nobody wakes up *dead* and I've got to find out how this came about. Of course, since you wrote the letter, I presumed it was not written from the *"other side"* and you must still be alive, but that didn't stop me from reading what you wrote word for word. Dream or not, it's a great story.

During my two years of therapy I've done a lot of dream work on myself and discovered dreams allow a direct view into the unconscious mind. They're usually in response to and an attempt to rationalize or bring about understanding of a recent thought *(last taste of Listerine = death)* that has not had enough attention paid to it.

What was probably on your unconscious mind was *death* and it pulled a fast one on you and slipped out in a dream. The thought that this might be your last use of *Listerine* to your conscious mind was a surprise. Death wasn't anything with which your conscious mind wanted to deal, and so it gave you a dream to think about. Your dream was dramatically laid out on the table *(In your face!)* so you could tell me *(and others)* and bring about a solution. Of course, I have a rule about dreams and that is, *only the dreamer can interpret his or her own dream*. If I'm wrong, so be it! I just bring it up to let you know the way I interpret my own dreams.

In any case, regarding writing, I'm encouraging you to write things down. If it's dreams, call it

DREAMS by *Charles Hazlewood.* Your dreams are so graphic they'd be wonderful about which to write and read. I wrote down my own dreams when I was in therapy and found out a lot about myself. Or write about stories that happened in your own life. When you have enough of them, you can list them chronologically and call them an autobiography or *Memoirs.*

My artwork pictures called *Reflections* were taken in downtown areas, like the city of Santa Monica or the sidewalk outside commercial plate glass windows. They are shot with my *Canon* digital $600.00 *Rebel* camera. It's the cheapest Canon camera that still will take all the attachments such as telescopic and wide-angle lens. It takes excellent pictures, clear and distinct, when enlarged up to four-feet by five-feet. If I choose only to use a detail twelve inches square, I can do so without losing any pictorial quality. Though many good double-exposure photos might be taken and used for artwork, my camera is not built especially for that advantage.

Shards are pieces of fender-bender. I go for walks on the Kanan highway a few times a week and sometimes notice where a guy rammed his headlight into another car leaving shiny pieces of broken headlight, glass or metal lying around. I pick them up and put them in my pocket, and when I come home arrange them on an outdoor glass table before taking raw material photo-shots. Back in the computer I pulled up each raw material shot and tried to organize it as an artistic composition in color, contrast, and an over-all idea of something I consider rewarding.

Regarding my pacemaker: about ten years ago I began having partial fainting spells at my computer. I even had a few semi-fainting spells when I was driving. At the computer things would go black and my head would slump down almost hitting my keyboard before I regained consciousness and composure would return. Subsequent tests showed I had an irregular heartbeat of 50-beats a minute that would sometimes stop for three or more seconds. The pacemaker brings my heartbeat up to an unfailing and regular 62 beats per minute. If I walk up the stairs or walk fast and my heart rate increases, the pacemaker does nothing, but my natural heart compensates. With a faster heartbeat, the pacemaker doesn't need to help. In installing a pacemaker they put you almost out, but keep you minimally awake, cut a slot in your upper chest and stick two wires into your heart; one at the top and one at the bottom. Then the 1-1/4" square size by 3/16" deep unit is placed under the skin in your chest and connected to exposed wires extending out of the heart.

Pacemakers last for 5 to 7 years. When the battery runs down, they put you almost out again, then open up the slot in your chest and change the unit by connection of the new pacemaker wires to the remaining wires still projecting from your heart. It's a fairly simple, non-painful procedure and the results are fabulous. Blood flows to my brain and whether I want it to or not, keeps me thinking. I try to walk for 40-minutes about 5-times a week and do about 5 to 10-minutes of arm and body exercises. I can't jog because my hip joints hurt. It's too painful. When I walk, I walk through the pain. I'll be 86 years old next month and I'm just going to keep walking until I can't any more.

Regarding electronic books: I still love to see the covers and illustrations and enjoy turning the pages and smelling the books and making collections of particular books of which I'm fond. I can buy a book for a buck at the library and when finished, take it back. It's surprising how many good books I get as somebodies leftovers. I've heard good things about electronic books, but that part of the electronic age has never appealed to me. I prefer good old natural books with pages and pictures. Reading of any kind provides an adventure *that lies within the realm of my own ignorance.* To me, it's always exciting.

Obama: Sure, send along the DVD. I'll watch it, but I can't believe it's going to be different than the excess of discouraging Obama information my friends have been sending me for the past 6-years. Supposing it's true and I'm convinced? *The Acid Sea of Reality* dictates he'll be gone in 2016. So what's next? I want a talented leader in negotiation with an expansive sensibility that's able to talk to leaders of the entire planet and address concerns I have in the last few pages of the *Acid Sea*. There's no future in arguable information?

Rather than end on a debatable note, please know I loved your dream-story and would like to read more. You are an interesting character and the world needs your particular expression. So, write!

Sincerely,
Doug

11-29-13

Chuck,

I received your DVD and letter and read your *Epiphany*. I'm always overjoyed when anyone has an *epiphany*. Suddenly something of which I've been only moderately aware comes more brilliantly into focus and I finally understand at a more profound level it's true meaning. Now you know, *God was looking out for you!* The train barely missed you!

That reminds me of the title of a book I'm always thinking of that sometimes drives Marge nuts. *(I'm always thinking up titles of books.)* It's called, *God Within, God Without*.

It involves the great, all-encompassing creative force or God that began with the *"big bang"* and produced the entire universe with all its stars, moons, suns, galaxies and earth. Within that concept what exists contains all that *acts*; the *living*, like man, woman, animals, fish, plants and birds, and all that *cannot act:* the *un-living*, like *water, wind, snow, mountains* and *molten rocks*. In the book, though all that is created by God, only *living humans* can make things happen.

In case of the train, multiple things happened. The train *(un-living)* was going to continue regardless. *God without.*

Gates *(un-living)* were closed by mechanical message. *God without.*

Cars *(un-living)* were responsive only to actions of drivers. *God without.*

Driver *(living)* acted in first car by turning left. *God within.*

Three drivers *(living)* acted and followed first car. *God within.*

You *(living)* acted and drove to safety.

God within.

You were saved by multiple *Gods within* and not by multiple *Gods without.*

Tell me if I'm wrong.

Regarding, *Dreams From My Father.* I've set down my feelings about finding solutions to problems on page-103 of *Harold and The Acid*

Sea of Reality. In a *nutshell,* they need to be rationally discussed by people knowledgeable on all aspects of the subject. Like an effective architect who must see the whole enchilada and be bigger than the project. Rule: *All parties must see the whole and be bigger than the project.*

Though all through my 54-year career I've seen the whole and been bigger than the house project, I'm not bigger than *"what to do with the president, the country and therefore, the world."* At the present time the United States rates number one as the most powerful nation in the world, and I do have my world concerns as are stated in *Harold and the Acid Sea of Reality* beginning on page 203. It includes thoughts on the health of the planet, religious conflicts, nuclear war, the severe population explosion and flowers of any nation, the educational facilities including universities and advanced programs.

Intelligent discussions on anything are necessary between those who disagree. The ideal moderator would insist on rationality as one of the main watchwords. Like football games, a violator might be penalized five yards for being offside or fifteen yards for unsportsmanlike contact.

The Obama book called *Dreams From My Father* is a real button-pusher for those not in agreement. I agree that *Frank Marshall Davis* was probably the real father and that explains his culture regarding books and music and his passion for helping the middle and lower classes in any way he can. He gets it from his real father. I detect a strong underlying message for the film. A meta-message, if you will. I understand, too, that what I see as the meta-message might not be the same as that seen by others.

That being said, i consider myself a normal, middle-class person raised by normal, middle-class parents, in a world that contains whatever both good and bad. I don't have all the answers to everything. Certainly, I care about everything and have written those thoughts and concerns in *Harold and the Acid Sea of Reality*. Reading the book is the faster way to get whom I really am. It already contains a lot of answers to most of my hopes and desires.

Regarding lies by someone no less than the President of the United States disturbs many of my feelings about life; that Obama's presidency could be based on lies and the occurrence of his travesty or the miracle of his presidency: the effort by others to capitalize on the lie and all that follows, or bask in a feeling of fulfillment at having done something right: the truths of misinformation that went to fight against Obama or perhaps the positive results that still may occur.

The Joel Gilbert story certainly made me want to ask questions!

Did Joel Gilbert himself pay to have this film made?

If not, who paid to have this film made.

Is the fear of the country becoming *Stalin-Lenin Socialist* the only reason this film was made, or is this just an ingenious, or the best theoretical attempt to presumably straighten America out?

If Obama was developing programs to which Gilbert agreed, would Gilbert have made this film?

If this film discredits Obama, what does the writer-director Joel Gilbert want to happen?

Are we supposed to impeach Obama?

Does he want the governmental agencies to understand the lie and vote against Obama's Health Care, foreign policy, environmental policies and position on immigration?

Does he want or not want Obama to sign the Keystone pipeline agreement and promote Canadian and eventually American fracking?

Your DVD certainly provoked thought and forced me to examine my own positions, though many of them are written about in *Harold and the Acid Sea of Reality*. Since I've seen the DVD twice and feel I know it pretty well, I'm returning it. Thank you for sending it. It told me things I didn't know. I hope you and your family had a nice Thanksgiving.

> Best wishes,
> *Doug*

12-24-13

Chuck,

Glad you have a great deal of admiration for my accomplishments, design ability, architectural awards, technical mind and writing ability, and that you think my writing of *Harold and the Acid Sea of Reality* has shown an acknowledgment of *our leaders true biography!* The true biography of our leader, God, is, I take it, the *Universal Force*. Right? This new factual information, for me is not new. I'm surprised you thought so.

Who is the true leader, unless it's the *Universal Force*, where have I mentioned a true biography of him or her? How am I able to reverse a long-standing bias through understanding of new factual information? Whose bias are we talking about, mine or yours? And what is the *factual information* of which you refer?

Let me be clear. I am in the best sense of the word an atheist. I'm aware of a *Universal Force* that some refer to as God. If others define the *Universal Force* as God, I have no objection. I have not been mindless about God or the meaning of the term God may have to others. To make sure of what I believe about the personal term, God,

I want you to understand it to be nothing more than a basic personal viewpoint. I've given the idea of God and the afterlife a lot of thought. Check my writings or ask Marge.

Note: in the *Harold etc. book* I've noted I agree with psychologist, David Viscotts definition of belief, *"Belief is knowing something is so, whether or not it is."* and *"knowing is the truth from a certain perspective."* Those definitions get us off the hook. You can believe and know something and so can I. They don't have to be the same. If we see things differently that's our inalienable right.

In the chapter on page 195, *Life After Death Isn't so Bad if you Survive,* I couldn't be more explicit on where I stand on life after death. To explain or analyze what is meant by *"Principle"*, or specifically, *principle of mathematics*: I'd have to look it up in the dictionary. It defines it as *"a fundamental truth or doctrine on which others are based. Rules of conduct or ethical behavior."* Regarding mathematics, one of the many fundamental truths under certain perspectives is one plus one equals two, or four times six equals twenty-four, or the dimensions of a hypotenuse equals the sum of the squares of the other two sides, or E=mC2, etc. What are you getting at?

If *Frank Marshall Davis* is Obama's real father who *"tutored and guided Obama's reading as well as many other things,"* he also instilled in Obama a love of poetry, music, action and compassion for others. In trying to help those in dire straits who couldn't help themselves, he set an example devoting his life to Communism and Socialism, methods he thought at that time as their only available road. Communism and Socialism are gone, but the problem is still here and compassion is still required.

As Obama was growing up, he had to play the political game. *If you can't stand the shame, get out of the game.* Another scenario could be that the main purpose of his drive was to help those who could not help themselves and not to turn them directly into Communists and Socialists. Barack Obama, leader of the Communist Party? Barack Obama, leader of the Socialist Party? I don't think so! Jumping to the conclusion that automatically one would become a hundred-year-old-Communist or Socialist is a hard call and that would be *"name-calling"* of the worst order. For Joel Gilbert to link Obama with words *Stalin-Lenin-Socialist* slants the article in a disturbingly negative direction. The *meta-message* is to remind the *un*-thinking person of *Stalins* and *Lenin's* atrocities and the *Hitleresque* rule of carnage. Rule *(1): To discredit a person, link him irrevocably with something discreditable.*

The *Russian Stalinist* days of 1930 and today are totally different. The world population has exploded from 2 billion people to over 7 billion people, the achievements in world-wide communication like the internet and television, have changed the world. Trips to Mars and the Moon and the world addition of the multi-national space station, the Hubble Telescope and multibillion dollar 17-mile cyclotron at CERN make the world of 1930 Russian and the United States entirely different. The Russian population in 1927 was about 120 million people. It's now a little over 143 million. The United States in 1927 was about 1 million. Now it's over 23 million. The context of the world is different. *Process* only happens in *Context*. The context of 1927 is not the context of 2014. We are worlds apart. The *local* economy is now *world* economy.

About my confusing sentence ending in *(the*

truths of misinformation) It would be more correct if I'd said: the truths of information of *mis*-information used to promote or fight against Obama, or perhaps the positive or negative results that will probably occur. In other words, the proofs that his ideas work or do not work are not as yet completely answered.

I should point out that if you promise free food and health care for people they're going to vote for you, but also, I know that many people have more pride in themselves as honest, hard-working people. It's my view some are free loaders and some are not. The statistic that *everyone* is going to live off the fat of the lamb is not correct and it's not a fact yet. The project might be defeated economically. Who knows about Michele or Hillary? Time will tell. I'm not sure.

I think Joel Gillbert wanted to discredit Obama and this video was the way to do it. If I were a politician, I might have to think twice about what to reveal, or not to reveal: that is, if it makes no difference as to what I want to achieve. So, if he thinks he is not a liar or a cheat, but his viewpoints are good for the rest of the country and world, I think he should continue.

When I ask should we impeach Obama, my question was really to you. Is that what you want? If you do, who would replace him? Joe Biden? Chris Christie? Ron Paul, Paul Ryan? Newt Gingrich? Michele Bachman? Ted Cruz? Jeb Bush? Who else? I'm looking forward to someone who is a good world negotiator, who will bring a human dignity to America and has compassion for people in hard straights, who has a view of the whole world and all with it's countries taken into account, someone who can keep everyone out of wars; somebody I think is smart rather than someone I think has no vision. Those are my views. It's good to be thinking.

Sincerely,
Doug

1-11-14

Dear Chuck,

Got your letter of January 4th, 2014. Dilemmas, like pain, are good since they're warnings of things needing resolution. I like to think I'm one person and that both letters came from me. I didn't have a need for communication and rational discussion on Obama and I am, like you, not one of those people who are knowledgeable on all aspects of the subject.

I've seen the Gilbert video twice and feel I know the substance of his message, but not the *totality*. Davis was a poet and I assumed Davis liked music since *"Obama's mother shared poetry and music with Davis."* He certainly had a passion to help the lower black class - not the black and white middle class - and I say that what young adults often learn is not only from words. My father never said a word to me, but loved and hugged me, worked three jobs to make ends meet, loved my mother as a real lady and cared with real passion for all children. The most important things I learned from him were without him ever saying a word. I observed from his actions and saw him as a roll model. From the video, I understood it was not in Gilbert's purpose to explain other more complementary things. His job was to paint a disturbing picture. Surely you can't claim that all Davis taught him was Lenin, Stalin, Socialism, and Communism?

There's a humanitarian aspect to assuring food, housing and education to those less fortunate. Are we going to let them die? Isn't America supposed to welcome all aspects of humanity regardless of creed, color and religion? I'm a percentage type of guy and do see that all of the poor and downtrodden are not freeloaders. I just bought a book from Barnes & Noble by Joshua Greene called, *Moral Tribes*. I'm only a 150 pages into it and found he examines differences of opinions, such as *Me versus Us*, selfishness versus concern for others, and Us versus *Them*, *Our* interests and values versus *Theirs*. At the end of the book, he claims to have a solution. I'm interested to see what he proposes.

I just ran into a quote by Mark Twain, *"It ain't what you don't know that gets you into trouble, it's what you know for sure that just ain't so."* Perhaps with Obama as President, we have just been had. When I say I don't have unwavering beliefs, that doesn't mean *all* my beliefs are wavering. I have firm beliefs and wavering beliefs that if I thought enough about each, I might change my viewpoint.

I do, know, however, as deceased psychologist David Viscott claims that *"Belief is knowing something is so, whether or not it is."* We can't go through life without beliefs. The beliefs I have, as you've stated for yourself, are arrived at through my experiences in life; coming from a loving family, being an athlete with artistic inclinations, being a social dork, marrying without certainty, having kids, making a life practice of building houses, divorcing and remarrying, reading and writing books, being into artwork, etc.

I've had a great life! I would love to have my tombstone say, *"Here lies Doug who fostered progress within context and change according to circumstance."* I don't consider myself an immovable object or an irresistible force, I consider myself a moveable object and a resistible force. I like to think I'm virtuous by being pliant, flexible and, should the circumstances describe, be willing to change.

Your next paragraph is hard to react to. I've obviously disappointed you. If our friendship relies on whether or not I hate Obama, there's nothing I can do about it. I don't buy your bottom line on Obama. Some of what you and Gilbert have to say is true, but I hear underlying messages that force me to draw different conclusions. If we're going to be enemies, all you have to do is stop the political-religious discussions and I'll quit too. I don't enjoy arguing with you. At times I see you as an irresistible force.

1 - It's not a justification, but most politicians lie. Name some that don't - George Washington, Abraham Lincoln? *(George had slaves. Abe helped free them.)*

2 - Why wouldn't Romney reveal his tax returns and how did he make his money?

3 - Davis probably is the true father and according to the video he did try to help the blacks and poor. I'm sure he also taught Obama better things than Lenin, Stalin, Communism and Socialism. What were the agreeable reasons why Obama revered the man?

4 – In those days Communism and Socialism were popular and to some, were the only viable avenues to pursue for downtrodden groups. The U. S. had only a third of the people it has today

and the technological advances since have been horrendous.

5 - Obama did deny his preacher, Faragon, and said he objected to the language Faragon used. Any approval of Faragon's animosity at that time would have injured his Presidency, his career, and would have been no help to people of America.

6 - You and Gillbert call him a Communist and Socialist, and though some of his actions seem to point that way, I still recognize a humanitarians motive to have America join other advanced countries for the public's common good providing health care, food, lodging and education.

7 - I'd rather have someone negotiating with Islam who knew something about Islam rather than someone with little or no experience about Islam. Most Muslims are not fundamental terrorists. You catch more flies with honey than a flyswatter and it's necessary we have an understanding negotiator. If you've ever read *The Art of War,* its conclusion is that war is the *last* alternative.

Regarding Michael Chrichton and my phrase, *"when it bleeds, it leads,"* all I was suggesting is that Chrichton might have been bleeding a little to lead for the sale of his book. I'm not arguing that the way to move people is through fear. For fear, read Gore's book, *An Inconvenient Truth*. Rush Limbaugh, Beck and Riley have their ridiculous stories as well. By the way, what are you getting at when you use the word, *"principle?"* I think of principle all the time. I have principles in my architectural work. I have principles in life. Everyone has principles by which they live and they're as different as fingerprints. Regarding the statement, *suicide bombers die by principle. Hitler killed Jews on principle*. Those not Muslim's are Infidels and need to be killed, etc.

At heart, I'm a humanitarian. I don't think it's morally correct for people to live in 15,000 square foot houses while others are going hungry. I don't think it's fair for 5% of the population to squander millions while 95% of the people are working their butts off for a pittance, etc. etc. etc. And I don't mean to discourage your reading of anything, Michael Chrichton included. All knowledge is power. That's a principle about which to live. To say no one could change my opinion is to do me a great disservice. I'm changeable, but I need to know the facts and decide for myself. That's one of my own principles.

An agnostic holds the principle, *I don't know.* Religious people say, I believe in God. *(My question is who is right?)* As an atheist, one of my principles is that there is no higher someone of infinite goodness watching over and taking care of all living things. The collective unconscious is the best we can do. See National Geographic's, video and read *An Inconvenient Truth.* Is Gore to be ignored?

Best wishes,
Doug

2-4-14

Dear Chuck,

Here! Here! I agree with your first paragraph. We're not arguing, we're debating! We are each presenting our hypothesis for acceptance or rejection without recrimination. *This is evidence of an open mind.* One thing that strikes me about the Obama discussion is this: Obama is President of the United States, and as such wields a great deal of power, not only in this country, but also in the world. He doesn't, however, have *complete* control *(like a dictator)* over the United States or everybody in the world. Witness the recent government shutdown by the Republicans and the present lack of Congressional cooperation.

He does have, however, like Churchill or Thatcher, the power of a leader *(as yet not impeached)* elected by the people. As President, he has an agenda. You can say what you want about what he says he wants and listen to your "meta-message." *(Note: Meta-messages are different for everybody because everybody is different from another.)* What we all understand is *"in the eye of the beholder."*

Forget President/Smesident, for a minute and

let's listen to the larger meta-message in our ears. If I had my choice, leaving everything that exists in place, what would I prefer a President to do? If I were president I would have to face *(with a hostile Congress and friendly Senate)* a host of big issues with a wealth of big problems; foreign policy, preventing nuclear war, *(I.e.: Iran, Korea)* counteracting growing terrorists, trying to facilitate an agreement between Israel and the Palestinians, the Syrian dictatorship and now the Ukrainian civil war, the use of drones, *(objected to because of "lateral damage" the killing of innocent people and its result, the spreading dislike for the United States)*, provoking or not provoking a ground war, what to do with the illegal immigrants, general lack of employment, economic disparity *(between five and eight percent of the world population have more money than all the people in the rest of the world)* global warming, ocean pollution *(carbon dioxide falls over two thirds of the earth thereby polluting the oceans)*, our over-dependence on foreign oil, the need for clean energy, *(wind-water)* world overpopulation, melting ice-bergs, rampant extinction of animal species, obliteration of forests, etc.

If I dislike Obama and say, *"He's a liar,"* how

does that solve the more immediate problems? I hear my own strong meta-message. It says, *"I want to help solve all the problems mentioned above."* Then I ask myself, *"How can I do that?"* Each of the problems is dependent on so many others: it makes me want to give up and say, *"Oh! The heck with it!"* and go somewhere else to live my life! Each one of the items I have mentioned would have to be debated and the proper solution found. I want to do the proper thing to save the planet and allow as many people and animals and plants to live productive, creative lives - even considering the fact that all living things have to kill other living things to live.

What we could do is discuss each of the above problems and see what we can make of it. For instance, I want to do the right thing for all of the above and I want the United States President to lead America and do the right thing for himself and America. But we all have to agree on what the right thing is. That's the crux of the matter.

There will be arguments and discussions on the best way to do things, or whether they should be done at all, or whether they have to be done immediately. That's what Congress and the Senate are supposed to do - debate and discuss and act. Are they doing it? Yes and no! They're doing it according to the best the Capitalistic United States system can do. At each instant they are looking at the *The Acid Sea of Reality*.

The Muslims were Arabs originated from nomadic, warlike tribes traveling the deserts and inhabiting the Middle East. It's one section of the world that needs to grow and be accepted by the world community. My belief is that it will do so. The Muslim's treatment of women is atrocious and the Koran, being the word from their only God, gives permission to kill the Infidel. We are the targets! How wrong is that? A Muslim can't criticize his religion for fear he'll be labeled Infidel by another Muslim and be mortally attacked. In America, I happen to know some very fine, educated, peace loving Muslims whom I'd trust implicitly. Too bad a few have to be terrorists and ruin it for everybody.

It is going to be interesting to see how China evolves. The newer architecture over there is modern and overwhelming. China, with their money, is just beginning to lead the world. I watch the McLaughlin Group, five experts, the questioner, John McLaughlin and his four associates two Republicans and two Democrats.

I get a good mixture of the important American issues. It's on every Saturday at 6:30 PM on KCET in Los Angeles.

I'm glad you liked the State of the Union Speech. I liked it myself. Regarding kissing all the pretty women; tell me you wouldn't do it!

Write soon,
Doug

2-28-14

Dear Chuck,

I've *not* given a lot of thought to the problems of the presidency. To do so would be to methodically research air and water pollution, foreign policy, national and world economics, long term affects of immigration, massive extinction of animal species, *(still going on)* diminishing natural resources, *(oil and gas reserves)* fracking, the national economic situation, strip-mining, drones, nuclear proliferation, clear-cutting forests, overpopulation and so on.

You're certainly right. The ideal president should have expertise in all these problems, but I have to admit that's one heck of a lot of knowledge. And of course moral values have got to be critical in the making of decisions because they impact not only the nation but also the world. Mob leaders won't make good presidents. Good or bad backgrounds influence everybody in their later lives. That doesn't mean everyone continues to make the same mistakes. I married the wrong person and stayed with it for 25-years until I met Marge and corrected my course. My youngest daughter had an alcohol and drug problem and almost went to prison. She corrected her mistake and is 22-years sober, a professional registered nurse and a magnificent mom raising two stalwart, straight-minded boys. *(12 & 14)* They look like Roman soldiers, are smart and into surfing, baseball, and water polo. I love 'em to pieces.

People who've had difficult early experiences often make the best teachers. Obama shows this. He meets a group of fatherless boys from Chicago every year and relates to them how it feels to be raised without a father. He gives them hope and tells them what to do about it. This is a productive, learning experience for these boys. He's showing real empathy and not doing it just to win votes. We have to examine motivation. As architects, are we going to design good environments, or make lots of money? Buildings occur, though largely through two different motivations.

Thinking about it, I'm afraid we're both bystanders letting the world parade go by. You and I who are both adult vote, but other than donating money to our favorite causes, how much can each of us do? I can't change the Muslim religion or any religion by myself, or stop the terrorists, or single-handedly prevent nuclear proliferation by myself, or stop genocide, let alone stopping fracking, oil spills, etc., etc., etc. My philosophy is this: Own what I believe, express it, and let the world do what it's supposedly to do. I'm not Martin Luther King or Charles Darwin or Abraham Lincoln. I'm just a retired architect living out the rest of my life as peacefully as possible.

Of course there is nothing attractive about the Muslim's or any other religious group infiltrating and taking over our *(or any)* country. Parts of the Koran say it's either my way or death. I've read only a few pages in the beginning and end of the Koran to get an idea of what it's like. I discovered it repeats itself and decided to give each idea a number so when the idea was repeated, I could give it a number, skip it and find something I could use.

Allah is the only God. *(1)* Do what I say and when you die you'll go to heaven. *(2)* If you don't do what I say, you'll go to Hell. *(3)* - etc. In this way I was sure I could cut the 423-page book down to 223 pages. Maybe a smaller book would be more useful. The listed items are absolutely worthless in the present day. The Muslim religion is based on fear. Do what I say, or you die!

The Muslim *"infiltration"* that you mention is, to my way of thinking, fear of the bad things you and others think might happen. The American Muslims are not the mid-east Muslims. American Muslims can't help being influenced by a more affluent and educated way of life. I don't think American, European, Chinese, Australian, Asian, or African people will sit still for the mid-east Muslim life-style, let alone their American-modified religion. They'll say as I say, *"Are you out of your freaking mind?"* No reasonable American would join a Muslim church and most American's would object to having Mosques in the neighborhood.

Educational facilities like schools and Universities are the flowers of any nation. Europe, America, Japan, Korea, Australia, New Zealand and maybe Russia and the Chinese set the standard by having schools and thereby educated people. The Palestinians, Syrians, Somalians, Libyans, Egyptians, Jordanians, Afghanistan's, etc. have absolutely *zero flowers* in their nations. Aware of it or not, its their job to catch up to the real world by providing jobs and advanced schooling. It's important they read more than one book. That's what libraries are for. For a better education the oil-rich Muslim in countries like *Iran* and *Saudi Arabia* send their young men out of the country to be educated in Europe or America.

Regarding your Internet photograph entitled, *"Another Jihadi Idiot,"* if I look carefully at the young man's face I feel compassion. I see someone on earth who's already in Hell. I see a totally destroyed, lost and abandoned human being and I can't help feeling sorry for him. Does that mean I condone his arbitrary murder of innocent men, women and children. Of course not! This poor, ignorant son-of-a-? must be prevented from his not-so-smart decision. Where are the flowers of his nation? Where was his loving family? Where was his Schools and University or meaningful career? In the mideast there isn't anything worthy for multitudes of these men to do. The mid-east countries have a tough job.

Christian Science: I have a close friend, Gene Grounds, who's a *Christian Scientist*. It's worked out well for him. He's 6-feet-4-inches tall, married, has two sons, one 6 ft. 6 inches tall and the other 7 ft. 0 inches tall. He and his wife live in a beach house in Oahu and have a catamaran. He's almost 80 and swims yearly in the 3-mile ocean marathon and wins as he also does with racing his catamaran. My close friendship with Gene is the full extent of my *Christian Science* knowledge. I don't think he's ever been to the dentist.

I didn't know the word *principle* had a special meaning to *Christian Scientists*. Having read your three *sentinels*, I now know the importance of the word, *principle*. I looked up *Christian Science* in Wikipedia and besides the normal definitions, found an article by Charles I. Ohrenstein, C.S.B of Syracuse, a lecturer in Boston, Massachusetts, who I quote, *"Principle means that which is ultimate, basic; that which is first or primordial, therefore the first cause"* and.." *Principle like God must be and is the primary, the ultimate consciousness, or Mind."* And.." *that God or Principle are one;* and.." *Mind, therefore, must be and is God, or Principle, the primordial cause of all."* *Principle* is also related to Mind, Spirit, Life, *Principle*, Truth or Reality." *let that mind be in you, which is also in Christ, Jesus."*

I take that definition to mean *what we think is what we are.* I notice you are a hate Obama guy. Never having given *Christian Science* thought, the nine-page article was truly enlightening. *"The greatest adventures lie within the realm of my own personal ignorance."* Truthfully, I feel a little less ignorant than I did before I read the article. After reading the Sentinel articles you sent, these nine pages and a few on Mary Baker Eddy. I have a better understanding of *Christian Science*.

> Thanks and my best to you,
> *Doug*

3-27-14

Dear Chuck,

Got your letter of March 23, 2014 wherein *"April first came in a little early this year."* I'm sorry you didn't get the five thousand, two hundred and eighty-eight dollars. If you'd done so, I'd probably asked you for a *"fiver 'til Saturday."* Glad nothing worse happened, like *"identity loss."*

On the Sheryl scene, I can't believe I was as good a draftsman as you've said. In San Diego I lost my job partially because my drafting wasn't good enough. I had about five or six jobs before I lost my last one in Pasadena and decided to get married. I drafted half of a whole Los Angeles high school during nine months of critical psychotherapy because I was convinced I married the wrong person. Twenty-five years later with the same marriage and after raising three daughters, I realized my first impressions were correct. That's when I met Marge. She and I have been together since 1976. I still think she's the cat's meow and still love her to pieces. She also has three daughters and one son. We have seven offspring between us with eight grandchildren ranging in age from eleven years old to twenty-two. Two have graduated from college. Well, I don't know how I got off on that kick. Anyway, until I see the drawings you'll send, I'm not sure they're really mine. I'll let you know.

How were Groves buildings? Did you like them? Was he God's answer to Frank Lloyd Wright? Did his buildings kick old Buddha's Gong? To answer your questions about those who worked for Groves, I do remember Weiderspan and a guy I think was the head of the office besides Groves whose name was Paul. Could that be Paul Mayberry? Also, I remember the pornographic guy who worked part time for Groves. His drawings spoke to parts of myself I had tried to ignore for 22-years. I'm trying to make up for it now.

Regarding Muslims. As a big group they are not the most educated in the world, however, I've known a few in California that were first rate. By that I mean sensitive to others, empathetic, not given to selling their religion and easy with whom to get along. Most Muslims who've been raised and educated in America and speaking English their whole lives and taking part in American activities such as attending school, work, entertainment are not the same as the uneducated masses of Muslims in Somalia, Kenya and the fundamentalists and terrorists in the middle east.

The mid-east Muslims need to read more than one book. I never lump all large groups under one idea or label them and try to remind myself that people are as different as fingerprints. I'm a percentage kind of a guy and try to base my beliefs and decisions on percentages and statistics. XX numbers of Muslims do *not* create happiness, but some *do*. XX numbers of Christians do *not* create happiness, but some *do*. XX numbers of Atheists do *not* create happiness, but some *do*. To decide on something, it's important for me to include the percentages and statistical numbers.

The division between church and state is a good rule by which educated nations need to live. I don't want any religion to dictate its special rules to an American like me in a supposedly free country who might disagree. As a personal statement: Politics is one thing, Religion is another. All voices should be heard. When you say:

"I'm not sure after one of your letters whether you are being sarcastic or you truly believe, as an entire body of hateful upbringing, they can only be influenced by Christianity."

1. By the phrase, *"entire body of hateful upbringing"* It sounds like you are saying *all* Muslims have a hateful upbringing. Many have a loving, educated upbringing, especially in the United States. In fact, that's why they come here - to escape the debilitating influence of people and leaders in their own country. *(2)* When you use the word *"they"* I infer you mean *all* Muslims. *All* Muslims are not the same, but as again are as different as fingerprints. *(3)* When you use the word *"Christianity"* does that mean *all* people should be Christians? Am I to understand you are saying *all* people should be Christians, thereby eliminating Buddhists, Jews, Indians, Muslims, Scientists, Agnostics, Atheists, etc.? A Muslim is never going to become a Christian, just as a Christian is never going to become a Muslim. The word, *infiltration,* used in the negative sense is a word of fear, a fact I have to recognize, and then make a plan. *(See enclosed essay.)*

I, too, am tired of accommodating the *less educated, less productive, complaining, unmotivated welfare receptionists.* People who are less educated need to be more educated. If *less productive* means they are less educated and sometimes *complaining*, who wouldn't complain if they had no education or job? *Unmotivated* is caused by a life without a place to go and that requires *education.* Therefore, with no education, few would be motivated. Hence, *welfare recipients!* With no welfare, the uneducated, unproductive, complaining and unmotivated, people including women and children would *die.* Are the government and I sure we want that? Education facilities are the flower of any nation. Education is what they and we need to create jobs. Education equals life. Lack of education equals death.

Thanks for your kind words on my minimal delving into the religion of *Christian Science.* Certainly, I was rewarded by my own slogan, *"the greatest adventures lie within the realm of my own ignorance."* At least I had one more adventure. Also, thank you for continuing our discussions. I'm a skeptic by nature and seize on everything that doesn't exactly ring true. Your letters are rewards to me because to answer them allows me to clarify my thoughts. I did get the articles on Groves and I *do* agree with you, we did *not* fully appreciate *whom* he was. 'Til next time,

Sincerely,
Doug

7-14-14

Dear Chuck,

Good to hear from you! Everything's OK here! My draftsmen and me have just been busy. I'm not overly *ticked* at you. Perhaps, just moderately. Hope that's OK with you and it will not affect our long friendship. After all, you're still the great person you always were, even though I think some of your presidential theories are misdirected.

Regarding drafting for Groves: I did receive drawings via Gmail, but they were so large as to be impossible to view as a whole. Perhaps a person with more computer knowledge could have made the Gmail smaller, but alas, not me. They remain faded back somewhere in Gmail archives. The parts I did see looked great and I even surprised myself I was that good. I must have been trying too hard with desperation for approval. On April 7, 2014, I wrote Sheryl Johnson a letter asking her to reduce the drawings to printable size, but never heard from her again.

Regarding Grove's domes: Eugene G. and I never said a word to each other. He had no idea who I was, and like me I had no idea who *I* was either. Our communication added up to a complete *"0"*. His domes I would speculate, might have been formed by joists and plywood in curves beneath, reinforcing steel, then spayed on top with gunite or shot-crete. Whether reinforcing steel could have been properly shaped without forms, I don't know. If it could, it would be cheaper in steel and forms would be unnecessary.

Regarding Obama: I like Obama. He's not perfect, but then I have to consider how much the Republicans thwart his good ideas. I listen regularly to the McLaughlin Group at 6:30 to 7:00 PM every Saturday on KCET. McLaughlin is Republican, Pat Robertson is Democrat, with Ellen Clift and two politically knowledgeable guests being neutral. They informally answer questions raised by McLaughlin about world issues. Last week he had hard questions on immigration, the Palestinian-Israeli conflict, attacks from the north on Baghdad and the importance of world climate change.

Regarding immigration: Obama is trying to act in the face of a sudden crisis. Why a *sudden* crisis of children, not from Mexico, but from

Central America? The answer is desperation and love for children within families. Raising children in countries with poverty and violence is insane, so the risk of a journey north to the United States is less than the risk of staying home. Illegal and immoral cartels are now making money by creating a pipeline to move hundreds of children to the United States. After they're here, it's no longer the cartels problem.

If they'd known, in retrospect perhaps the U. S. could have stopped the busses. This tangled ball of spaghetti was handed to the U. S. on a platter. Now entering the picture are humanitarian considerations ignored by the cartel. Whether Democrats or Republicans are in control, dealing with thousands of children is not going to be comfortable, easy or cheap. I can hear lots of complaints, but Boehner's speech on *"sending them safely back to their parents"* evokes my response, *"Ha! Ha! Ha!"* Processing costs money and delivery back to Central America ain't cheap or safe. While I mourn, I have to let the world be itself. Obama is doing his best to act. If he didn't do something *now*, he'd be criticized for not acting on time. The perfect solution, instead of fixing the clutch, it's time for the problem to go away.

I ran across a philosophic gem by Mark Twain. *"It ain't what you don't know that gets you into trouble, it's what you know for sure that just ain't so."*

That means whatever I *"know for sure"* should be taken with a grain of salt. Like everybody else, I have thoughts, hopes, desires, intelligence *(or lack thereof)* and motivation. It's better to be aware of the *whole* than aware of only a part of the whole and this philosophy applies to life as well.

Sandwiched, as I am, with my own brand of talent, passion and motivation, it's necessary I know as much of the *whole* as I'm able. The whole of being an architect involves *(1)* the client, *(2)* the site, *(3)* the contractor, *(4)* the Building Department and *(5)* the clients and my own desires, reason for living, enthusiasms, creative ability, who I am in the world and knowing the *whole* as much as I'm able. I don't get into trouble because I *know something for sure*.

Regarding George Will's article entitled *Stopping a Lawless President*: The article brings to mind one faction of the Muslim's fighting against another faction of the Muslims. So far there's no bloodshed in America, but it seems the U. S. has the same issue as the Middle East, Republicans against Democrats, or one faction of the U. S. fighting another faction of the U. S. It's only a matter of degree, of course, but if Marge and I didn't see eye-to-eye we'd be just like the Muslims and the U. S., we couldn't get anything done. Where's America's unified team? Amanda, who's raising two magnificent young boys at 12 and 14, often *catches them doing something right!* As tiny as this little philosophy is it's her prime source of action. I'd like to see the benefits exaggerated once in a while rather than the negatives and have both factions give credit where credit is due.

These days are art days for Marge and me. Marge has five remarkable PrismaColor Pencil drawings now showing at the Hillcrest Center for the Arts in Thousand Oaks, CA. She's been invited to show with a dozen other artists usually working in oils or watercolors. We'll be attending a reception on August 11 from four to six PM. I just picked up two of my own artworks from a photographic show in Ojai last month. I was pleased to get an Honorable Mention on one called *Throes of the Dance*. I'll also be delivering four abstract photographs to the Ventura County

Art Association in Ventura, CA. I'll enclose a few to show you. The reception will be this Friday. We enjoy the receptions. It's the only social activity we have going these days. Also, I'm doing cards and selling them in art stores for $5.00 each. There's no money in it, but what else is new? See enclosed. Write!

> Take care, my old and dear friend,
> *Doug*

6-15-15

Dear Chuck,

I was glad to get your letter a few minutes ago. For exercise in the late afternoon Marge usually goes for a half-hour walk and returning, checks the mailbox and brings up the mail. Most mail asks for money, but occasionally we are pleased to get a letter from close friends, like yours. I like snail mail better than Gmail. There's something more personal about getting an actual page handled and typed by a real person. I enjoyed the example of the clear camera focus you sent me by Gmail. I was blown away by its clarity.

The subject of illegal immigrants triggers a complex moral issue. Americans have to decide whether we want to share our wealth and education with our less fortunate humans *(brothers and sisters on this planet)* along with the burden of new costs and inconvenience, or live with the cost and inconvenience of keeping them out by a wall, and holding camps and prisons that will include the price of food, care and lodging until we can send them back from where they came. The more convenient and less costly solution might be to *(1)* stop them from coming in and *(2)* make Americans out of the ones already here. Making them Americans would mean teaching them the language, giving them food, a temporary living place and keeping them healthy while they assimilate our present customs. That should take place within a generation or two. The cost and inconvenience of getting rid of them might be more than the cost and inconvenience of assimilating them and it certainly is morally correct.

Assimilation means within a generation or two they will know English, have jobs and be paying taxes. I always root for the long-term solution rather than short-term. Better minds than mine are working on the problem. This is the best I can do with limited information.

Regarding your note on me being *agnostic* and having a *mind set* because "*daddy said so*". Let me explain; It's my understanding an agnostic is one who questions the existence of God. I presume that means the God or Gods of the world as described in books and bibles of Jews, Christians, Moslems, Witnesses, Mormons, Catholics, Christian Scientists, etc. Of course, I don't believe in any of those Gods. You must understand I do believe in many of the teachings of these religions, such as the ten commandments, the idea of love, treating every living thing as if it was holy, the golden rule, etc., and in general, all the ideas of love and man's loving devotion to all things on the planet *(except snakes and snails and puppy-dog tails)* including immigrants. I'd like it better if God were understood as the *Universal Force;* the whole cosmos as initiated by the big *"bang."* God = *Universal Force.* OK!

What you said in your letter, *"It's impossible for a man to learn what he thinks he already knows"*, that philosophy could also describe a man unwilling to keep an open mind. It's certainly possible a man with an open mind can correct old thinking with new knowledge. Whenever I come up with what I think is true and set it down for all to read, I'd like to add, *"Tell me where I am wrong."* If someone tells me where I am wrong, I examine it in light of the knowledge I already have and it either changes the way I see things or solidifies my former knowledge. I find this question is a cool way to keep my mind open. To my way of thinking, mind-open = life and mind-closed = death. I put it in these two terms because, as my lawyer stepson would say, it keeps me *"going in the right direction"*,

My dad was not a father whose authority must be obeyed and which, if I'd complied for whatever reason, I would have had *"faith"* because he said so. My dad disowned his own authority and threw the ball in my court. He gave me freedom and permission by saying *I could figure it out for myself.* That is, I could examine religions, philosophies, cosmic theories, Darwin and all things in the world, including religions and make up my own mind. He did not hand me *"faith"* like in theories recommended by Catholicism, Judaism, Mormonism, Moslemism, etc. He did not promote that kind of *"faith."* In effect, he said keep an open mind, examine everything pertinent and decide for yourself. He did not insist upon any pre-established belief. I don't have a mind-set. My mind is never fixed in one place. Should I understand something new with which I agree. I change!

Regarding explosions and gunfire: I didn't hear about those on TV. We don't get Los Angeles newspapers. Your experience must have been a little unnerving. I'm glad no one was hurt and I hope they caught the guy. I hope the homeowner had insurance.

Regarding reading: I heartily recommend

doing it! Reading 28 books in 5 months are 5.6 books per month. That's outstanding! Either you're a pretty fast reader or the books are really short.

For pleasure I've lately been reading a few W. Somerset Maugham books; *Liza of Lambeth, The Magician, The Theater, The Gentleman in the Parlour, The Hero, Christmas Holiday* and *The Complete Stories of Somerset Maugham*. For entertainment, try old Somerset, he's an exciting writer. To learn something go on Amazon and buy used books for less than the shipping fee. I've gotten several for one penny plus $3.50 for shipping. Try *The Sixth Extinction*, by Elizabeth Kolbert and *Moral Tribes*, by Joshua Greene. *Why Does the World Exist?* by Jim Holt. *The Hockey Stick* and the *Climate Wars* by Michael E. Mann. Others I'd recommend are *The Eerie Silence*, by Paul Davies, *The Feeling of What Happens* by Antonio Dimassio, and *A New History of Life* by Peter Ward and Tom Kirschvink.

Thank you for your good words on leaving a *"heritage"* of work of which I'm proud. I am fortunate enough to still have passion and motivation to do writing and artwork. I did not get a Rolex or a golden umbrella at my retirement. The only people that were with me were one called Marge. We went to a local Italian restaurant called *SPRUZZO'S* and had spumoni.

I'm still doing abstract photography and showing in local galleries. In the past week, I got an Honorable Mention at the Ventura County Art Council gallery and a second place in the Conejo Valley Art Association in Newbury Park. I'm getting quite a list of awards after 5 years of work and 80 shows. Marge and I are hiring a person to help us print books. Marge has 20 PrismaColor pencil drawings she wants to record in a book called *In My Mind's Eye* and I have a two more books to print, one is a 12 inch x 12 inch comic book called, *A Book About Everyday Stuff* with Jorge' and Merle, *(Jorge's a grown dog and poet. Merle is an adolescent pup)* A more serious book I hope to print will either be called *A Tale of Two Houses* or *Trial by Fire*. It's about building my first Malibu Pedestal House, its burning and building a new, more fire-resistant house over the same foundations. I'll send you copies. I should have them some time this year.

Best wishes and thanks for writing. Keep reading! Reading = life. Not reading = death.

More knowledge = life. Less knowledge = death. Finding solutions = life. Complaining with no action. = Death. More love = life. Less love = death. Tell me where I'm wrong!

'Til next time!
Doug

5-4-15

Dear Chuck,

    Thanks for writing. Regarding money for artwork: there isn't any. You might ask what does it cost me to do artwork? In the past five years I've sold two pictures for $300.00 each that cost me about a hundred dollars to produce excluding 30% for the gallery. I also sold three other pictures for $100.00 each with the gallery getting 30%. I designed and did the backing for the piece I sent you with 6 individual frames. I worked on the frame on my office drafting table and a workbench out the back door. I save about half the money framing them myself.

    Brian, my printer over the past 5 years, has become a friend and is a sort of partner in crime. I deliver him a CD with my photo and size. He prints it with his five-foot wide printer. His machine will print anything on paper five-feet wide up to a hundred feet long. Another device he uses is a five foot wide *"cold press"* similar to the 1920's electrically operated hand-driven clothes ringer. He then mounts the positive paper with a colored artwork print to black Gaterboard, a harder more dimensionally stable foam-board. He then runs the Gaterboard and colored print through the cold press a second time laminating a thin piece of clear plastic over the top of the colored photographic artwork. The clear plastic furnishes an archival finish inhibiting the colored print from fading in direct sunshine. It also allows safely wiping the colored artwork with a damp rag. Brian's cost is about 10-cents a square inch. A photographic artwork 30"X30", or 900 sq. in. would cost about $90.00. If the final print is 20"x20" sq. in., or 400 sq. in. it would cost $40.00.

Art, music, writing, dance, poetry and the rest of the arts are bad ways to make money. Architecture isn't much better. So why do I do it? Because I want to, can afford it and enjoy meeting friends and companions at openings and sharing my artwork with other artists. Marge is a brilliant PrismaColor Pencil artist. She had a one-woman show with 23 pieces. It cost her a month's rent of *($1,000.00,)* for the studio and she sold 5 pieces for about $1,000.00 a piece, for a total of $5,000.00. Framing 23 pieces cost $10,000.00. If you do the math, you can see it's a financially losing proposition. I find that while in retirement, I have to do something with my life and I think pursuing the classics, art and writing is nevertheless worthwhile.

Considering *"life and death philosophy"* and *"the fine line between opposites"*, if I go the present way I think = life, if I go the opposite way, what's true is If I do not think = death. It makes sense to do something that's, I think = life, than the opposite. In managing old age the continual message is *follow what the Universe is doing. That* tells me to keep moving. To move = life. Not to move = death. So I encourage myself to *"keep moving"* and the next question is *"what do I keep moving at?"* I happen to have the motivation and some talent, a dharma or two if you will, to make things such that, when I'm done I can look back at *"me"* and remind myself I'm alive and a worthwhile person. I can say, *"I did this!"* and that *"I feel good"*.

You know the thrill you got when you designed the best of your great houses? When you finished it, it was real. It had three dimensions. You could kick it and it hurt your toe. It existed because you put in your effort and got it built in accordance with your spirit and intelligence. When it was done it looked back at you and told you, *"By God, Chuck, you should be proud of yourself, you're damn good!"* It told you good things about yourself which allows you to say to yourself, *"I like myself."* = life. Contrarily, if you say, *"I don't like myself"*, that = death. The point is, because I've lived 87 years and finally have the right to say my piece, it's important I set myself up now for something I'll enjoy later = life. Not to set myself up for something I'll not enjoy later = death.

I have a female friend who is helping me print my books. She's 50, has a 20-year-old daughter in college and a 17-year-old son who's a senior in high school. She also has a 56-year-old husband who can't work, but must stay home because he has dementia and sleep apnea. She cares for him and their house and is the sole breadwinner. I asked her how she manages all that? She said she'd developed a philosophy of living in the now. That is, living the fullest life that is possible for her to live, enjoying fresh air, trees, clouds, her work, her dog, her family, her artwork, her writing, her publishing and her time with the kids and seeing all the joyful things that happen in the *"now"*. Her idea is that if she made all the *"now's"* happy without dwelling on the things that were unhappy and over which she had no control, she'd eventually remember a happy life.

We all have an autobiographical memory. We all can remember the bad things or the good things as if they were now. I'm not suggesting I don't deal with the bad things in life. If I have the power, I certainly have to deal with them. But I'm suggesting that I should enjoy the good moments while they're here. *"Take joy in the good moments."* = life. *"Do not take joy in the good moments."* = death.

Regarding immigration: It's my philosophy that *"all decisions should be made within the context of what else is occurring."* To be the best in architecture, I need to know the whole: wind direction, sun, rain, snow, where the views are, the geology, the neighborhood, the location of stores and schools, how far it is for the owner to go to work, the personal program of the owner, the budget, the building department regulations and tract rules, soils, gas, water, electricity, what's legal and what's illegal, the contractors with all the professions of their subs, my own draftsmen or lack thereof, then the financing and what's happening with the country, plus the owner's and my own lifelong intention, etc. etc. etc. In other words, to do a good job, I have to know the *"whole"*.

Therefore, for a problem as complex as immigration, legality or illegality is only part of the problem. Of course, immigrants have entered the country illegally! Everyone agrees on that. The question for debate is how to handle it, if not from a moral standpoint then from a mercenary and annoyance standpoint. If I'm unable to understand the whole, it's like expecting the car to run without wheels.

Your second paragraph is nothing more than a complaint. I agree with your second paragraph. If you have a new idea no one has thought of, certainly you should bring it up for consideration. Stating the problem over and over annoys everyone and wastes time. Yes, it's illegal, yes it's costly, yes it's not fair, yes it's extremely bothersome, yes we don't want to pay for it, and yes we shouldn't have to deal with it. On the other hand the problem still needs a solution. How will this happen and who's going to do it?

Marge, for 3 years has had Parkinson's disease. She didn't ask for it. She's just got it. It's costly. It bothers us. It's not fair. She can't get a good night's sleep. She's always tired. My life is and will be as her caretaker. I'm not complaining, but the question is, what we are to do about her having Parkinson's disease?

To answer the immigration question, America is going to do all of which it is capable, which might mean sending would-be immigrants back to where they came from, or turn them loose on the other side of the border without food, medical care, or education, which, no doubt, would turn them into even more desperate people leading to more dishonest ways for they and their children to stay alive. Or should we just accept them and educate them and do our best to make them responsible Americans? I should think this has been given lots of thought by those in charge, and whatever is happening is probably what *should* be happening. It is all I can do to encourage the thinkers to do their best. I'm afraid the leaders of America operate with the prejudices of a property owner's association, but I can't do anything about the bad parts of that.

Regarding God = the Universal Force principal: I could agree that if something or other created the Universal force that might be called God, it reinforces the notion that reasoning without facts leads to false conclusions. Thinking it out for myself and after reading extensively on the subject, I think there are some things we poor humans can never know; at least at this stage of our evolution. I do know, and so do some very important thinker's on the subject like astronomers and scientists know the Universe was not created by some supernatural power. If you say you know, that's your faith.

Faith is different from knowing. According to Psychotherapist, David Viscott, *"Faith is knowing something is so, whether it is or not.* The job of the skeptic is to keep an open mind and examine every contribution for a possible answer. I'm not sure whether it makes any difference whether someone named God began the whole Universe or not. We have arrived here on the only planet we know of, alone and stuck with no possibility of escape and lost in the cosmos with overpopulation, melting icebergs, diminishing forests, pollution of atmosphere and ocean, with global warming and extinction looking us dead in the face. Did a personage called God create this? It is apparently true that only the human species is unlikely to fix it. That's the long-term view and the one in which I agree. Long-term = life. Short-term = death.

Regarding books: Reading books = life. Not reading books = death. My son-in-law, Chris Lewi, a lawyer, sometimes makes a worthy suggestion. He will say, *"If you're reading books, you're going in the right direction!"* Of course, what you read will have an impact on what you remember and do or not do. What do you want to remember? So it depends on whether you're reading for enjoyment, or to study a subject, or reading because you like humor, or about science, philosophy, history, poetry, art or just a good story, etc.

Charles Krauthammer's, subjects sound interesting. I'd be interested in what Chuck Hazlewood is reading. You get positive points from me if you read anything. *Of course, that and five dollars will get you a cup of coffee.* One book I enjoyed was *The View from the Center of the Universe* by Joel R. Primack and Nancy Ellen Mabrams, Copyright 2006. It moves along quite easily and presents a powerful, optimistic picture of the Earth within its own context - *the Universe.* He said architecture is a way to im-prove culture in any society. That depends on people of the society. Thanks for your provocative thoughts.

Writing thoughts = life. Not writing thoughts = death.

Best wishes,
Doug

1-5-18

Chuck,

    I got your letter today responding to Marge's passing. Thank you for your kind words. It's been over a year since her death, but it's been well over three years since she was truly healthy. About 4 years ago following diabetes, she rather quickly lost 6 pounds and the doctor diagnosed her with lymphoma. Over the next year she had 6 expensive shots at about 1-1/2 month intervals. Parkinson's disease then followed and she began tremors, or light shaking. She took 5 pills a day for a year and a half for Parkinson's that more-or-less controlled her shakes. It was mentally and physically wearing until she was diagnosed with pancreatic cancer. Her pancreatic cancer turned out to be the slow, cruel kind because it kept her alive for 8 painful months instead of the usual 3-1/2 painful months.

    With cancer, in the space of 3 months she had 6 operations on her liver, pancreas and gall bladder. During the last part of her illness, and while she was lying in a hospital bed in the living room, we had full time hospice care that kept her comfortable. While her 4 kids and my 3 kids were around and concerned, we gave her fluids and helped her eat as much as possible.

Over a period of 8-months, her body finally refused to accept either food or water. She lost weight, eventually weighing about 80 pounds before becoming fatally hungry and dehydrated. It was hard to see her starved and dehydrated to death by tiny increments while there was nothing I as caretaker or anyone could do. Marge handled it bravely, stoically and maintained a reasonable self throughout. I couldn't help but love her.

Well before her extended illness and while hiking through Malibu's Charm Lee Park, Marge was healthy and thinking ahead, we discussed the idea of having a memorial bench when we died overlooking the ocean in western Malibu. After her death, I arranged for that to happen. We now have a steel bench that takes about 30-minutes to reach by walking through Charm Lee Ranch along the hiking paths to the top of a steep bluff overlooking the ocean.

The bench was supported on concrete foundations by two of my favorite contractors, John Wiley and son, Jamie. The memorial bench overlooks the broad Pacific Ocean and Point Dume upon which Marge raised her 4 kids. Off the broad, blue horizon of the Pacific can be seen Catalina Island and off to the west, a few Santa Barbara Islands. We had a bench ceremony on Marge's birthday, October 2nd, 2017, when she would have been 84 commemorating her life with her close lovingly family attending.

The copper plaque in the center of the six-foot wide back of the steel bench facing the ocean says, *(on top)* REMEMBERING *(next line)* DOUG RUCKER – 1927 – *(blank for date) (next line)* MARJORY KRON LEWI-RUCKER – 1933 – 2016 - *(last line)* WITH LOVE. The tribute went well and we felt good about the memorial bench. I've returned three of four times and have noticed hikers resting on our bench and enjoying the view. This gave me a good feeling. The hikers seemed happy after their 30-minute hike being able to sit down and enjoy the magnificent view.

I didn't know what life would be like alone after 42-years with Marge as my constant companion, but here it is over a year later. On the 31st of December 2017 I had my kids and Marge's kids and some 30 people over to my house to celebrate my 90th birthday.

Since living alone, I appreciate my eldest daughter, Viveka, an Acupuncturist and my youngest daughter, Amanda, a registered home health nurse for Kaiser Permanente. Amanda, with 2 teen-age sons, Nathan and Christopher. They have been calling me every night to make sure I'm OK. Viveka watches my finances and walks with me about once a month. Lilianne, my middle daughter, is also loving and attentive, but lives in Windsor an hour or two drive North of San Francisco with her three kids and PHD husband, Ernie.

To keep myself occupied, I'm reprinting formerly written books in hard cover and having them listed on Amazon. Before I die, I just want to have my say. My stepson says I'm leaving a paper trail. I think I've sold three books and made $15.00 each, so as you can see it's not exactly a moneymaking deal. I'm working on a new picture book about my own mid-century houses called *Building a Home That Loves You*.

In the book, I'm taking what people like in life and incorporating whatever that is into their houses. Now it means I have to figure out what people like in life. So far, it's harmony and integrity. Maybe you have some ideas. I have to admit I'm quietly proud I have one house *(out of the 90 houses and 50 or more remodels)* that has been designated as a Los Angeles *Historical Cultural Landmark* along with the biggies, Wright, Schindler, Neutra, Lautner, Quincy Jones, Ellwood, Abell, Drake, Gehry, etc.

Then, I've been staying in touch with my artwork. Yesterday I finished entering a show sponsored by the *Thousand Oaks Art Association.* The whole show is a hundred and fifty or so pieces with a hundred or more artists competing for $2,400.00 worth of prizes in six or so categories. My category is *digital photography*. I've just sent in 4 submissions, two of which will probably be accepted.

So that's the news from Lake Wobegon. What's happening with Chuck? Write and fill me in. I've been out of touch for many years. I enjoyed your response about Marge's death. For 42-years, she was the light of my life and I can never complain that during my life I wasn't in love.

With my best wishes,
Doug

Other Books by Doug Rucker

Personal Journey
    Poems predicting next phase of life.

Early Stories
    Autobiography Birth - University.

Groundwork
    Autobiography - Marriage & office.

Growing Edge
    Autobiography — Office & Recreation.

Moving Through
    Poems & "No Think" pastels.

Book of Words
    Short Stories - Humor & philosophy.

Harold & the Acid Sea of Reality
    Thoughts on fantasy & reality.

Trial by Fire
    Burning & rebuilding personal home.

Building a Home that Loves You
    Architectural Philosophy - pictures.

Transitions
    Realism, Reflections & Abstract.

Thinking in the Abstract
    Deciphering abstract art.

Poetries
    Poems - Abstract Art/ Poetry Prose.

Brief Biography

Born on the last day of 1927 and after finishing eighth grade in Chicago Illinois, Doug was awarded a scholarship to the Chicago Art Institute before entering Chicago's largest Austin High School. In football, swimming and track he was awarded 7 letters while pursuing a 3-year course in architecture. At the University of Illinois between 1945 and 1950, he completed his Architectural Bachelor of Science degree, and thereafter worked as a draftsman in Denver, San Diego and Pasadena. There, he married his first wife, Karon and soon after, became a licensed architect and had 3 marvelous daughters, Viveka, Lilianne and Amanda, while working as an important designer/draftsman in a Brentwood Village architectural firm. He designed and built their first house that his family lived in for about 5 years. In 1966 he built another Malibu dream house, a *"House of the Heart"*, the main floor floating on a 26-foot square pedestal, 40-feet high with wrap-around decks and spectacular views of the ocean, shore line, Surfrider's Beach and 65 acres of wildly blooming geranium gardens. He received much newspaper and magazine notoriety before it was burned to the ground in 1970 by a devastating brush fire. By 1972 he'd built another more fire-resistant *"House of the Head"* over the same foundations. It can be found in the, Architectural Guidebook to Los Angeles by David Gebhard and Robert Winter. It was similarly honored, but lost to a divorce in 1980. His new wife, Marge, and he enjoyed 36 years of creative life in a very small house he designed on an acre of land in the mountains above Point Dume in the County of Los Angeles. Marge died of various illnesses in December, 2016. Now retired at the age 97, Doug lives alone cared for by his three daughters, Viveka, Lilianne, Amanda, son-in-law Tom Rincker and Minerva his part-time caretaker and her dog.

www.ingramcontent.com/pod-product-compliance
Lightning Source LLC
Chambersburg PA
CBHW040732220426
43209CB00087B/1611